Books in the
Alpha Wolves of Yellowstone Series
by Rick McIntyre

———•———

The Rise of Wolf 8:
Witnessing the Triumph of
Yellowstone's Underdog

———•———

The Reign of Wolf 21:
The Saga of Yellowstone's
Legendary Druid Pack

———•———

The Redemption of Wolf 302:
From Renegade to
Yellowstone Alpha Male

Praise for *The Rise of Wolf 8*

"[McIntyre] spins the best stories about wolves that anyone will ever tell, ever."

DOUGLAS W. SMITH, senior wildlife biologist and project leader for the Yellowstone Gray Wolf Restoration Project

"Rick McIntyre's book, *The Rise of Wolf 8*, has no match in literature. He presents the personal lives of wild wolves in a riveting narrative without equal for its detail and insight. This work took decades of devotion and consumed most of his life. We are so fortunate that we can, through this book, share the ride."

ROLF PETERSON, Michigan Technological University, author of *The Wolves of Isle Royale: A Broken Balance*

"Rick McIntyre knows more about Yellowstone's wolves than anyone living or dead. This book is a must-read and a treasure for anyone who loves the places where wolves howl."

THOMAS D. MANGELSEN, nature photographer and conservationist

"To follow the ever-changing destinies of the Yellowstone wolves is to witness a real-life drama, complete with acts of bravery, tragedy, sacrifice, and heroism."

JIM AND JAMIE DUTCHER, founders of Living with Wolves

"The stories of the different pack members are reminiscent of Ernest Thompson Seton's animal characters, but these wolves were and are from real life, and what they reveal will become a classic study in animal behavior."

BERND HEINRICH, professor emeritus of biology at the University of Vermont and author of *Mind of the Raven*

"Rick McIntyre is a fabulous researcher. He dedicated his life to documenting the histories of generations of wolves in Yellowstone. I envy the hours he has spent in the field talking to people about wolves and tracking the wolves' movements in the most beautiful country in the world."

SCOTT FRAZIER, director of Project Indigenous, Crow/Santee

"Yellowstone's resident wolf guru Rick McIntyre has been many things to many people: an expert tracker for the park's biologists, an indefatigable roadside interpreter for visitors, and an invaluable consultant to countless chronicles of the park's wolves—including my own. But he is first and foremost a storyteller whose encyclopedic knowledge of Yellowstone's wolf reintroduction project—now in its 25th year—is unparalleled."

NATE BLAKESLEE, author of *American Wolf*

"This book clearly demonstrates that these apex predators are an essential ingredient for maintaining the integrity of the diverse ecosystems in which they live. *The Rise of Wolf 8* is a must-read—one to which I will return many times—for anyone interested in wolves and the natural world. Wolves (and humans) are lucky to have Rick McIntyre."

MARC BEKOFF, PhD, University of Colorado (Boulder), author of *Rewilding Our Hearts: Building Pathways of Compassion and Coexistence* and *Canine Confidential: Why Dogs Do What They Do.*

"Rick McIntyre knows and understands the wolves in much the same way that a traditional Native would. He knows their birthdates and birthplaces, their family and their family history, their personalities, preferences, strengths and weaknesses, their character or lack thereof. In other words, he knows them as individuals—and not just as subjects for scientific study. He KNOWS them."

JOHN POTTER, wildlife artist, Anishinabe

RICK MCINTYRE
Foreword by ROBERT REDFORD

THE
Rise of
Wolf 8

WITNESSING
THE TRIUMPH OF
YELLOWSTONE'S
UNDERDOG

GREYSTONE BOOKS
Vancouver/Berkeley/London

Greystone Books Ltd.
greystonebooks.com

Cataloguing data available from Library and Archives Canada
ISBN 978-1-77164-780-9 (pbk)
ISBN 978-1-77164-521-8 (cloth)
ISBN 978-1-77164-522-5 (epub)

Editing by Jane Billinghurst
Copyediting by Heather Wood
Map of Northern Yellowstone National Park by Kira Cassidy
Cover and text design by Nayeli Jimenez
Cover photograph of a gray wolf by Jim Cumming
Printed and bound in Canada on FSC® certified paper at Friesens.
The FSC® label means that materials used for the product have been
responsibly sourced.

This book was written after the author finished working for the National
Park Service. Nothing in the writing is intended or should be interpreted
as expressing or representing the official policy or positions of the US
government or any government departments or agencies.

Greystone Books thanks the Canada Council for the Arts, the British
Columbia Arts Council, the Province of British Columbia through the
Book Publishing Tax Credit, and the Government of Canada for
supporting our publishing activities.

Canada

Greystone Books gratefully acknowledges the xʷməθkʷəy̓əm (Musqueam),
Sḵwx̱wú7mesh (Squamish), and səl̓ílwətaʔɬ (Tsleil-Waututh) peoples on
whose land our Vancouver head office is located.

"There is no fundamental difference between man and the higher mammals in their mental faculties. . . . Animals, like man, manifestly feel pleasure and pain, happiness and misery. Happiness is never better exhibited than by young animals, such as puppies . . . when playing together, like our own children."

CHARLES DARWIN, *THE DESCENT OF MAN* (1871)

"[Your studies of] chimpanzees made us see them as individuals and have empathy for them."

STEPHEN COLBERT TO JANE GOODALL (2014)

CONTENTS

PRINCIPAL WOLVES

—————

THESE FAMILY TREES cover the numbered wolves in the packs that feature most prominently in this book: Crystal Creek, Rose Creek, and Druid Peak. Numbers in triangles indicate male wolves and numbers in circles indicate female wolves.

Crystal Creek Pack

The alpha pair, alpha male 4 and alpha female 5, arrived in Yellowstone in January 1995, along with their four male pups. The family was from Alberta.

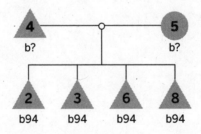

Rose Creek Pack

A mother and daughter, 9 and 7, arrived in Yellowstone in January 1995, and were introduced to lone male wolf 10. The three wolves were all from Alberta. 9 and 10 formed a pair bond, and 7 dispersed to become the alpha female of the first new pack to be formed in Yellowstone: the Leopold pack. Wolf 2 from the Crystal Creek pack joined her to become the Leopold pack's first alpha male.

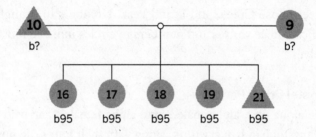

In addition to the five pups identified by number in the family tree above, the litter of 1995 included three more male pups.

Druid Peak Pack

A mother, 39, and her three daughters, 40, 41, and 42, arrived in Yellowstone in January 1996 and were introduced to lone male wolf 38. All five wolves were from British Columbia. Later that year, the five Druids were joined by a young male, wolf 31, who is thought to have come from the same pack in British Columbia as their alpha male, 38. Although 38 and 39 were the alpha pair, they did not produce any pups together. Wolf 38 did produce pups with two of 39's daughters, 41 and 42.

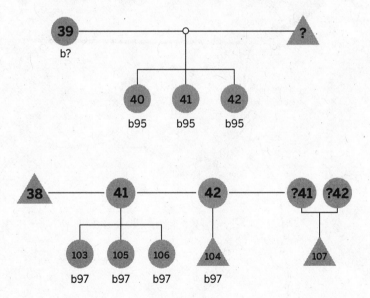

Wolf 163 was born into this pack in 1998. His mother was likely 40. To find out who his father was, you will need to read on.

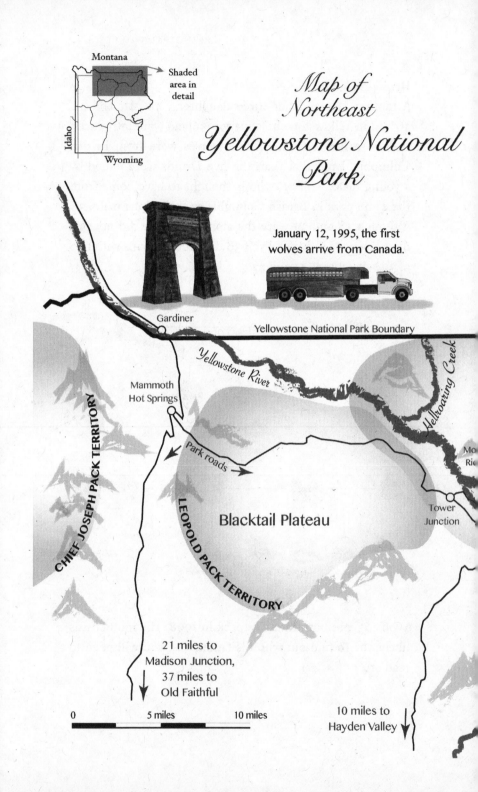

Map of Northeast **Yellowstone National Park**

January 12, 1995, the first wolves arrive from Canada.

Montana

Shaded area in detail

Idaho

Wyoming

Gardiner

Yellowstone National Park Boundary

Yellowstone River

Mellowroaring Creek

Mammoth Hot Springs

CHIEF JOSEPH PACK TERRITORY

Park roads

Blacktail Plateau

LEOPOLD PACK TERRITORY

Mo
Rio

Tower Junction

21 miles to Madison Junction, 37 miles to Old Faithful

0   5 miles   10 miles

10 miles to Hayden Valley

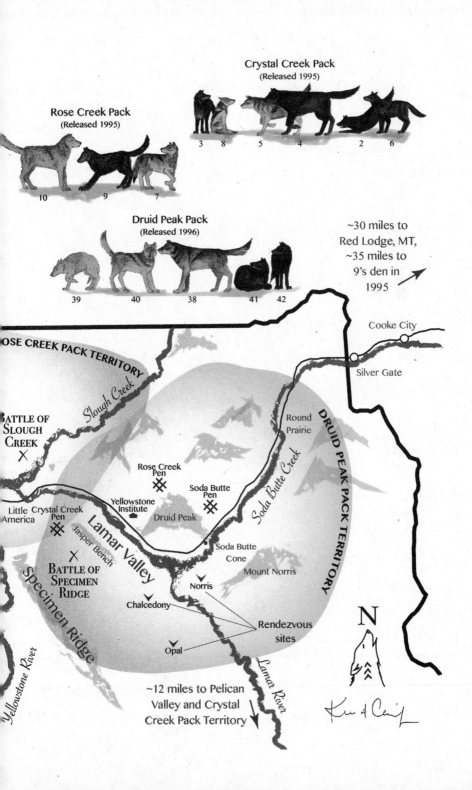

Crystal Creek Pack
(Released 1995)

3 8 5 4 2 6

Rose Creek Pack
(Released 1995)

10 9 7

Druid Peak Pack
(Released 1996)

39 40 38 41 42

~30 miles to
Red Lodge, MT,
~35 miles to
9's den in
1995

Cooke City

Silver Gate

OSE CREEK PACK TERRITORY

Slough Creek

ATTLE OF
SLOUGH
CREEK
X

Round
Prairie

DRUID PEAK PACK TERRITORY

Rose Creek
Pen

Soda Butte
Pen

Soda Butte Creek

Yellowstone
Institute

Druid Peak

Little Crystal Creek
America Pen

Jasper Bench

Lamar Valley

Soda Butte
Cone

Mount Norris

BATTLE OF
SPECIMEN
RIDGE
X

Norris

Specimen Ridge

Chalcedony

Rendezvous
sites

N

Opal

Yellowstone River

Lamar River

~12 miles to Pelican
Valley and Crystal
Creek Pack Territory

FOREWORD

A MERICA'S WILDNESS HAS always fed our souls and inspired our dreams. For many, wolves are the undisputed icons of nature, independence, and freedom. For others, wolves are considered a threat to livestock, their families, and their future.

Writer and biologist Rick McIntyre has a compelling story to tell. It begins in 1926, when park rangers shot the last of these apex predators in Yellowstone. Few people at the time mourned the loss.

Yet as wolves continued a sharp decline throughout the US (eventually landing them on the threatened and endangered species list), a movement took hold, and thirty-one wolves were eventually reintroduced back into the park in the mid-1990s. Decades later this bold wildlife restoration is considered the most successful ever undertaken.

McIntyre picks up the tale from there, sharing his journey of passion and dedication, adventure and perseverance, as he observed these packs returning to their native lands. He has spent those years hiking into the backcountry, filling thousands of pages with meticulous notes, and setting up roadside scopes for travelers from all over the world who have come to see and learn more about these creatures.

In particular wolf 8, one of the very first to roam free, captures his attention and his heart, and grows into the main character in this narrative.

Through McIntyre's eyes, we witness and learn about wolves as unique individuals living with breathtaking intensity—and it's impossible not to be awed by their loyalties to each other, keen intelligence, and will to survive.

Given this intimate portrait, and the controversy that continues to surround wolves, we're left to wonder how to balance their important role in the ecosystem with the interests of people whose lives they cross when they step beyond the protection of the park.

There are no easy answers, but I believe also that there is no limit to human ingenuity if we are truly seeking solutions. Information and data are vital, but so are stories that help us empathize with wolves. Both can inspire and inform decisions for the future. This terrific book presents them side by side and gives us the chance to decide for ourselves... itself a sacred act of American freedom.

ROBERT REDFORD
Sundance, Utah

PROLOGUE

T HE STORY TOLD in this book is an epic one, filled with heroes and heroines who struggle to survive and defend their families. A story that includes all the elements of a great tale: warfare, betrayal, murder, bravery, compassion, empathy, loyalty—and an unexpected hero. It is a story that deserves to be told by a literary genius such as Shakespeare, Homer, or Dickens. None of those writers was available.

If Shakespeare had written a play about these characters and their lives, he might have invented a prologue set at a wolf den, deep within a forest. The scene might have looked something like this. Three male pups, all jet black, run out of the den and begin playing in a meadow. Each of the black pups looks robust and strong. Then a fourth pup tumbles out, the smallest of the litter. This pup looks totally different from his brothers for he is a dull gray color, more like a coyote than a wolf.

Then a huge black wolf strides onto the scene. This is the pack's alpha male, the father of the four pups. The black coats of the three larger brothers indicate that they are going to look just like their father, and their stoutness implies they will someday equal or even surpass him in size and strength.

The gray pup, it is clear, will never look anything like his father and will never be as impressive.

Shakespeare might then have had a narrator voice a prophecy:

Three of these young sons will become mighty alpha males, control vast territories, and have many sons and daughters.

Looking at the three husky black pups, it is easy to imagine which of the four sons will become successful alpha males. Then the narrator adds:

But one son will die young and in great disgrace.

At that point the undersized gray pup trips over his own legs and falls on his face in the dust.

Finally, the narrator makes one final, cryptic pronouncement:

One of these sons is destined to be considered greater than the greatest wolf who ever lived.

This book will tell the story of two wolves: the greatest wolf who ever lived and the one that was greater than him.

1

Wolf Interpreter

THE FOUR MALE pups described in the prologue were born in the spring of 1994 in Alberta, Canada, east of Jasper National Park. Their family was known locally as the Petite Lake pack. At that time, I was in the extreme southern part of the United States, working as a seasonal naturalist for the National Park Service at Big Bend National Park in west Texas, the most remote national park in the lower forty-eight states. As I drove toward an abandoned Depression-era ranch near the Rio Grande river, where I was to lead a tour for park visitors on local history, I tried to figure out how to get around a major setback in my life.

The previous day I had received a call from Tom Tankersley, the assistant chief naturalist at Yellowstone National Park. We had an understanding that I would have a job that spring at Yellowstone as the park's wolf interpreter, and I would specialize in giving talks about the possibility of reintroducing wolves into the park. I would be the world's only official wolf interpreter. But Tom had called to tell me that government funding had not come through for this new position, and the offer would have to be rescinded. He was sorry, but it could not be helped.

As I continued driving through the desert landscape, I tried to come up with a plan to save that position. I passionately supported reintroducing wolves into Yellowstone, and I felt with my previous experience with wolves while working for the National Park Service in Alaska that I could help win acceptance for the proposal. Beyond that, I had a gut feeling that I just had to be there. A door had been closed, and I had to find another way through.

An inspiration suddenly came to me. I led the walk, then rushed home and called Tom. I had a proposition: What if the position could be privately funded? After a few moments of silence, Tom said he would check. He called back the next day to say there did not seem to be any regulations prohibiting private funding, so my idea might work. He gave me an estimate of how much would be needed for the four-month position and a deadline for having the money in an account managed by the Yellowstone Association, the nonprofit that handled donations for the park.

After thanking Tom and finishing the call, the reality of my situation set in. How were we going to get that much funding? By my standards at the time, it was a lot. Fortunately, I was about to leave on a lecture tour to publicize my recent book, *A Society of Wolves*, and I would be speaking to several large groups in California. The timing was ideal.

It turned out that I was lousy at explaining the situation. I had trouble clearly stating why Yellowstone needed funding for the wolf interpreter position. In my first talk, in a Southern California community, I mumbled a few words about how people could support Yellowstone's wolf reintroduction proposal. If anyone wanted to help, they could speak to me after the program. No one did. Then I drove north to San Francisco to give a talk at the California Academy of Sciences. The deadline Tom had given me was a few days away. I knew this would be my largest audience, and if my plan was going to work, it had to happen there or I would have to give up on going to Yellowstone.

Four hundred and fifty people showed up. I managed to make a slightly more coherent explanation of the Yellowstone wolf story and how people could help. After the program, a crowd gathered around me, and several people asked about funding the position. A few made small pledges. I was grateful for their contributions, but as I added them up in my head, I knew that I was nowhere near the goal I needed to reach.

A young couple stood there quietly, listening to what I was saying. I sensed they had to leave. The man stepped forward, handed me his business card, and said that they would like to help. He told me to look at the back of the card. I did and saw that he had pledged $12.50. I thanked him and said I would let him know when I reached my goal. Other people asked me wolf questions. As the couple walked off, I took another look at the back of that card and saw my mistake. The pledge was not for $12.50. It was for $1,250. With previous pledges, that amount would bring me close to the figure Tom needed.

That was the moment I knew the Yellowstone job was going to happen. I excused myself from the people gathered around me and ran off to find that couple. They were still nearby. Feeling somewhat awkward, I asked if I had read the back of the card correctly. The man, Gary, modestly told me I had. He introduced me to his companion, Trish. We talked about wolves and Yellowstone, and I thanked them for their generosity. The next day I called Tom and told him we had enough funding for the job. We set a starting date and began to plan how the new position would be structured.

I packed up, left Big Bend the first week of May, and started to drive north toward Yellowstone. It would be a trip of about fifteen hundred miles, and I planned on doing it in three days. As I drove through hundreds of miles of barren west Texas countryside toward New Mexico, I had plenty of time to think about how events in my life had led to this new assignment.

I WAS BORN in Lowell, Massachusetts, and spent my first ten years in the nearby small rural town of Billerica. We lived in a renovated schoolhouse on Concord Road. A farm was across the street and the surrounding land was full of woods, ponds, brooks, and fields. It was wild country, the perfect place for a kid like me. Looking back, I feel I had an idyllic childhood there.

It was the 1950s and, to use a modern term, we were free-range kids. On summer days and weekends during the school year I would go wherever I wanted in the outdoors, either alone or with other guys in the neighborhood. Some days it was fishing in one of the local ponds, other days it was just walking in the woods. Sometimes it was biking along the endless back roads in our town. The common thread was being outdoors, and as I spent more and more time out there, I became increasingly fascinated with wildlife. I was drawn to the small fish in the brook behind our house and would occasionally catch some of them and keep them in a small aquarium. I found turtles even more intriguing, and I put a lot of thought and experimentation into figuring out how to catch them. After examining one, I always released it.

I recently heard astrophysicist Neil deGrasse Tyson say, "All kids are scientists." That triggered a memory of something I did back then. The farm across the street had two dogs: Rex and Shepy. Like all farm dogs, they were never tied up and did whatever they wanted. I noticed that most mornings Shepy would walk off into the woods and come back late in the day, just like I often did. I wondered what he was doing out there, so one morning when I saw him set out, I followed him and watched as he wandered through the woods and fields, investigating various scent trails. He was exploring the country, much as I had been doing. We were kindred spirits. That day was a preview of how I would eventually study wolves in Alaska and Yellowstone many decades later.

After finishing college as a forestry major at the University of Massachusetts, I got a job in Alaska at Mount McKinley National Park and was stationed sixty-six miles out on the park road at Eielson Visitor Center. The mountain was eventually renamed Denali, its original Native American name, and the park's name was changed, as well, but when I arrived there they were both still called McKinley. Eielson was built on the Arctic tundra at a location with a commanding view of the 20,320-foot then-Mount McKinley, the highest peak in North America. The interior of the visitor center had displays on mountaineering, geology, wildlife, and tundra vegetation, but the view was the main attraction. My duties included conducting nature walks out on the tundra, leading half-day hikes, and giving evening campfire talks at Wonder Lake Campground. That campground offered a close view of McKinley and was one of the most spectacular viewpoints in the entire national park system.

Visitors to McKinley were always especially taken with the park's large population of grizzlies, and I felt the same way. That first summer I watched every bear I saw until it went out of sight. I was also fascinated by the other animals in the park: caribou, moose, Dall sheep, and the vast array of migrating birds that nested on the tundra. But from my first day in Alaska, my goal was to see a wolf. At that time in McKinley, they were rarely seen.

One day at Eielson I heard people talk about seeing two wolves stalking a mother moose and her two calves. When my shift was over, I drove out and found the cow and her calves. Then in the nearby willows I spotted two gray wolves, the first wild wolves I had ever seen. They moved back and forth in the brush, trying to get between the cow and her calves. When the wolves gave up on the calves and went out of sight, I headed back to Eielson, elated at having had the experience of seeing those wolves.

The next summer I returned to Alaska, and I kept going back for a total of fifteen summers. In 1975, the Alaska Legislature asked the federal government to officially change the name of the park to Denali, and although that was not to happen until 1980, Denali was the name commonly used in

Alaska at that time, and the name I will use for the park and the mountain from now on.

Wolves gradually became more commonly seen, and I spent a lot of time watching and learning about them. I read Adolph Murie's 1944 groundbreaking book, *The Wolves of Mount McKinley*, and eventually found a high viewpoint where I could watch the distant den site of the East Fork pack, the pack Murie had studied in the late 1930s and early 1940s. I got to know the pack's alpha female and her mate, the alpha male, a wolf with a bad limp. I saw the alphas and the subordinate adults in the pack take care of the pups at the den site, and I watched as they hunted caribou and dealt with grizzly bears that came too close to the den. I once witnessed the pups stalk their sleeping father and leap on him like they were attacking a prey animal. He shook them off good-naturedly, walked off, and resumed his nap.

During those years of my life, I was essentially a migrant worker. I was in Denali for those fifteen summers and, in the winters, had jobs in desert national parks such as Death Valley and Joshua Tree in California. I switched to Glacier National Park in 1991 and spent three summers there. My third year, I worked in the Polebridge section of the park, the prime area for seeing wolves. Wolves had been killed off long ago in Montana and other western states, but in the late 1970s a few crossed the border from Alberta and settled in the northwestern part of Glacier, the first wolf recolonization in the American West. The wolves were hard to spot because of the thick forests, but I saw a number of them that summer, including a family playing in a meadow.

Around that time, I was asked to write a book about wolves. I had built up a lot of wolf sightings in Denali and Glacier, and I had read all the books written about wolves and most of the scientific papers on them as well. I knew the biggest current issue was the possible reintroduction of wolves into Yellowstone. They were native to the area when Yellowstone was set aside as the world's first national park in 1872, but the early park rangers, like nearly everyone else in the country at the time, felt wolves were no good. In 1926, they killed the last few wolves in the park.

For my book, I made several trips to Yellowstone to interview the Park Service biologists and managers planning the potential reintroduction. They included John Varley, director of the Yellowstone Center for Resources, and Norm Bishop, a longtime Yellowstone interpretive ranger doing public outreach in communities near the park. After that I spoke with many of the advocates for reintroduction, such as Hank Fischer with Defenders of Wildlife and Renee Askins of the Wolf Fund. I also went to Helena, Montana, to interview Ed Bangs, the US Fish and Wildlife Service biologist who was the wolf recovery coordinator for the Northern Rockies. While I was there, I went to a hearing on the reintroduction and testified in favor of it.

By the time my book *A Society of Wolves: National Parks and the Battle over the Wolf* was published in the fall of 1993, I was well versed on the Yellowstone wolf reintroduction proposal. In addition, I had been around wild wolves in Alaska and Montana for sixteen years. All that led to my going to Yellowstone in the spring of 1994 to start my job as the park's wolf interpreter.

WHEN I ARRIVED in the park in early May, I met with Tom Tankersley, and we went over the schedule he had created for this brand-new position. I would live in a government trailer at Tower Junction and present wolf programs throughout the two-million-acre park. I had already developed a slide show on wolves and the possibility of bringing them back to Yellowstone. I would give that program weekly at the Madison and Bridge Bay Campgrounds and occasionally at the Mammoth and Fishing Bridge Campgrounds, as well. Those programs would go through the end of the tourist season in early September.

I would also give daytime talks in several of the park visitor centers. For those programs, I would show video footage of wolves in Denali that my friend Bob Landis had taken during the years I worked there. Bob would go on to shoot numerous wildlife documentaries for the National Geographic television channel and the PBS program *Nature*, including many on the wolves of Yellowstone. The plan that summer was that I would put Bob's footage up on a monitor, describe the behavior of the wolves, and then talk about the proposed wolf reintroduction in Yellowstone.

Most of my time would be devoted to what the Park Service calls "roving interpretation." That involves going to where the biggest crowds are and informally talking to as many people as possible. Only a small percentage of park visitors go to scheduled ranger programs, so there is a need to reach the many people who do not attend any of our talks. It is like being a street preacher, rather than a minister giving a sermon in church.

I borrowed a wolf pelt for the summer and tried to figure out how to attract the attention of visitors so I could tell them the reintroduction story. I remember the first time I stopped in the parking lot at Tower Fall. I put on my park ranger hat, checked my uniform, took out the wolf pelt, and started to walk toward a crowd of people. I was immediately surrounded by scores of visitors who all asked about the pelt.

I developed a talk where in a few minutes I explained that wolves were native to Yellowstone but had been killed off by the early rangers. The Park Service later realized what a mistake we had made and now hoped to reintroduce wolves to the park by bringing animals in from Canada. I talked to one cluster of people, then moved on to the next group. I could get my message out to about three hundred people an hour that way. Most of them would never have gone to any of the park's formal programs. To add variety to my work, I went into gift shops with the pelt and strolled through the aisles. As they had done in the parking lots, people rushed over to see what I was carrying. I then went through my short talk before switching over to the next aisle.

In midsummer we got word that the park's wolf reintroduction proposal had been approved by Bruce Babbitt, President Clinton's secretary of the interior. From that point on I revised my talks to say we would be bringing the wolves back during the coming winter. By the end of the season, I estimated that I had talked to over 25,000 park visitors about wolves and the plan to reintroduce them to Yellowstone.

That summer I finished work on my second wolf book, *War Against the Wolf: America's Campaign to Exterminate the Wolf*. It was a collection of historical documents going back to colonial days that traced the origins of anti-wolf bias in America and the reasons our country was determined to kill off all the wolves, even in national parks. The book also reprinted some of the early writings that began to portray wolves in a more positive light, such as Ernest Thompson Seton's *Lobo the King of Currumpaw*. It finished with the current plans to bring wolves back to the Northern Rockies and the Southwest. I commissioned wolf biologists and wolf advocates to write new articles for the book and wrote several new essays myself on wolf recovery and the plans for Yellowstone. The book came out in the spring of 1995.

My job as wolf interpreter ended in September, the traditional time when park visitation dropped off. I had been invited to go on a speaking tour of Ireland and England that fall, and I left the park to talk about wolves and the Yellowstone reintroduction in Belfast, London, and several other cities. One of my lectures was to the Royal Zoological Society. I also was interviewed several times by BBC radio stations. The wolf reintroduction program was capturing international attention.

That fall I thought a lot about the summer I had spent in Yellowstone. A comment made by Henry David Thoreau, who grew up just a few miles from where I did in Massachusetts, came to mind. He was born in 1817, much too late to see wolves as he walked through the woods of New England. In an 1856 journal entry, he expressed his sadness over the extermination of wolves and other native animals in his area. He felt he lived in a tamed and emasculated country. Thoreau spoke of the sounds and notes of the natural world no longer in his woods and mourned that he had to live in an incomplete land. He went on to say, "I listen to a concert in which so many parts are missing." The most prominent

2

Wolves Arrive in Yellowstone

AFTER MY TRIP overseas I returned to Big Bend for my second winter season. That fall Yellowstone hired two wolf biologists to plan the reintroduction and to monitor and research the wolves after their release: Mike Phillips and Doug Smith. Mike, the only biologist in the country with wolf reintroduction experience, was designated the project leader. He had worked for the US Fish and Wildlife Service as the Red Wolf Recovery Program coordinator and had overseen the reintroduction of that species in North Carolina.

Doug had spent many years researching wolves in Isle Royale National Park in Lake Superior under the supervision of Rolf Peterson of Michigan Tech. He later did wolf studies in northern Minnesota for the US Fish and Wildlife Service under Dave Mech, who was also a professor at the University of Minnesota. Dave had helped Durward Allen of

Purdue University start up the Isle Royale wolf study back in 1958. It became the longest-running wolf research project ever undertaken. Rolf took over supervision of the study in 1974. Doug's experience with Dave and Rolf meant that he was trained by two of the top wolf biologists in the world. Mike had worked on Isle Royale with Doug, so they knew each other.

In January of 1995, I went on another speaking tour, this time in Ohio. That happened to be the same time the wolves arrived in Yellowstone. Fourteen wild wolves, members of three packs and one lone male, had been captured in Alberta, about 550 miles north of Yellowstone. They arrived in the park in a horse trailer on January 12, and I watched the CNN television coverage of the event.

When I got back to Big Bend, I got more detailed information from friends in Yellowstone. The three packs had been put in separate acclimation pens in the northern section of the park near where I had been based at Tower Junction. That was an area with a high density of elk, the primary prey species for the original Yellowstone wolves, as well as for the packs newly arrived from Alberta. Each pen was about an acre in size.

The first two wolves captured were a mother and her female pup from the McLeod pack. Another female pup from that family had just been shot and all the other members of that pack, including the alpha male, likely were dead, killed by hunters and trappers. The mother (designated as wolf 9) and pup (wolf 7) were the only known survivors. They were placed in a pen built in Lamar Valley behind the Yellowstone Institute. Because the pack did not have a breeding male, a wolf captured as a lone male was also put in that pen. He was given the number 10. The group was named the Rose Creek pack.

The pack described in the prologue ended up in a pen six miles east of Tower Junction and became known as the Crystal Creek pack. The alpha pair (female 5 and male 4) and their four male pups were the wolves television crews filmed when rangers and other park staff unloaded their metal cages from the truck and took them up to the pen on a sled pulled by mules. The pup designated as wolf 8 was the little gray in the prologue. At 72 pounds, he was the smallest of the four brothers and of all the fourteen wolves brought in from Canada. The last one in his pack to be captured, he almost got left behind.

The five members of the Soda Butte pack, known in Canada as the Berland pack, ended up in a pen in Lamar Valley, east of the Yellowstone Institute.

As death threats had been made against the wolves, armed law-enforcement rangers guarded the packs twenty-four hours a day while they were in the pens. Those rangers, who trudged through deep snow in subzero temperatures throughout the long Wyoming winter nights, were the unsung heroes of the story. Thanks to their dedicated work, no wolves were harmed while in their pens.

During the ten weeks the wolves were in confinement, Park Service employees brought in elk, deer, and bison carcasses twice weekly to feed the packs. These were mostly animals killed in collisions on local highways. Wild wolves need an average of 10 pounds of meat per day in the winter to meet their energy demands. For the six wolves in the

Crystal Creek pen, that amounted to over 350 pounds per week.

On March 21, at 4:15 p.m., Mike Phillips and Steve Fritts of the US Fish and Wildlife Service opened the gate of the Crystal Creek pen, then quickly walked back to the road. Everyone expected the wolves would notice the open gate and immediately run out. But they stayed in the pen. There was only one gate, and since humans always entered the pen through that gate, the wolves were probably afraid of approaching it.

On March 23, the biologists cut a hole in the fence, well away from the gate, and placed a deer carcass just beyond that opening. The next day a monitoring device indicated wolves were going through the opening to the carcass but returning to the pen after feeding. The pack had not yet figured out that they were free to leave. On March 30, five of the six wolves permanently left the pen, and the final one joined them the next day. It was a ten-day process, but by March 31 all the Crystal Creek wolves were roaming freely in Yellowstone.

Park Service employees, park visitors, and local residents regularly saw the Crystal Creek wolves exploring the region around their pen. They were also seen playing together, a sign that the wolves were getting comfortable with their new home. Those were the first sightings of a free-roaming wolf pack in Yellowstone in sixty-nine years. The family stayed near the pen for the next four weeks, feeding on winter-killed elk and hunting elk on their own.

THE THREE WOLVES in the Rose Creek pen had a different story. The big male, wolf 10, who at 122 pounds was above-average size for an adult male, left that pen soon after the gate was opened on March 22. That made him the first of the wolves from Canada to leave a pen. The two females, like the Crystal Creek wolves, were more hesitant about departing. The mother wolf seemed afraid of the open gate and refused to approach it. We did not know it at the time, but she was pregnant. She and the male had formed a pair bond during their time in the pen.

The male stayed just outside the pen. That would have been very dangerous from his point of view, for he knew the area was the most likely place for humans to appear. But he loyally waited for his new mate and her daughter to come out. It would be like someone who had escaped from prison staying nearby and waiting for his friends to break out, regardless of guards patrolling the area and the high probability of being recaptured.

On March 23, the biologists hiked up toward the Rose Creek pen, intending to cut an opening in the back side of the fence, like they had just done at the other pen. A blizzard drastically cut visibility as they approached the site. Near the pen they heard a howl, then saw 10 staring at them, just fifty yards away. The crew, not wanting to disturb him, turned around and quickly hiked back down the trail. The male followed them downhill, howling continually as he escorted them away from his new family. He returned to the pen area once the people were gone. Sometime in the next day or two, the mother wolf left the pen with her daughter, and all three Rose Creek wolves began traveling throughout the surrounding area.

A hole was cut in the fence in the third pen on March 27, and the five Soda Butte wolves walked through it to a nearby deer carcass later that day. Like the Crystal Creek wolves, they returned to the pen at first, then left the site for good two days later.

All fourteen of the wolves brought down from Alberta were now exploring this new land that would become their home. The reintroduction team had successfully carried out the plan to bring wolves back to Yellowstone. Now it was up to those wolves to do the rest: restore the park to its original condition, including the sounds of howling wolves.

3

My First
Wolf Sighting

W HEN I FINISHED up in Big Bend in early May 1995
and began driving north toward Yellowstone, my goal
for the coming summer was to see at least one of the
newly released wolves. I knew the wolves brought down to
the park had been captured in areas of intensive wolf hunting and trapping. Mortality there due to humans was often
40 percent annually. That meant they had good reason to be
afraid of people and would want to avoid them. I hoped that
if I did a lot of backcountry hiking I might get lucky and get
a glimpse of a wolf. I would be living again at Tower Junction,
a few miles from where the Crystal Creek wolves had been
released. That would give me a better opportunity of seeing
one of them.

I reached the park's Northeast Entrance on the evening
of May 12 and drove west toward Lamar Valley. When I got
there, I saw Bob Landis, the wildlife documentary filmmaker

I had known in Denali, parked on the side of the road. He told me he had been filming the six Crystal Creek wolves, but they had slipped into the trees just before I arrived. I had missed them by minutes. I looked for them anyway, then, feeling discouraged, I drove on to Tower and moved into my trailer, thinking I had missed my one chance of seeing wolves that year.

I got up early the next morning and drove the ten miles east to Lamar. When I arrived at around 6:00 a.m., all the Crystal wolves were in plain sight about a half mile south of the road. I saw the black alpha male, the whitish alpha female, and their four young sons, now about a year old. The smallest wolf in the pack, the gray yearling, stood out, his dull fur in stark contrast to the sleek coats of his three bigger black brothers. My goal had been to see one wolf that summer and, on my first full day back in the park, I was watching six wolves roaming Lamar Valley.

As I observed the pack, several people pulled over and asked what I was looking at. I had the same spotting scope I had used to view the East Fork pack in Denali, and I let those visitors look through it at the distant wolves. When they saw the wolves, their faces shone with joy and excitement. More people stopped, and I helped them all see the wolves. Most of them commented on the beauty of the alpha female, the majestic appearance of the huge alpha male, and the gorgeous coats of the three black yearlings. No one said anything about the little gray, wolf 8.

The alpha female, wolf 5, marked an old elk carcass site with a squat urination and scratched the ground with her hind legs. As she walked off, the alpha male, wolf 4, came along and did a raised-leg urination over her scent mark, then scratched the site with both hind legs. Wolves have scent glands between their toe pads, so the scratching enhances the marking of their territory. Any other wolves coming along would understand that an alpha pair had marked this site. The Crystal wolves were claiming Lamar Valley as their own.

The female continued in the lead and the other five followed. As I was soon to learn, it is the alpha female who makes most of the decisions for the pack, such as choosing the direction of travel, and the rest of the wolves, including the alpha male, follow. The pack approached a big bison herd. Several of the bison glanced at the approaching wolves but showed no concern. Mature bull bison can weigh up to 2,000 pounds, and cows get up to 1,000 pounds. Since the average weight of adult wolves is around 100 pounds, bison are ten to twenty times heavier. Most of the bison did not even bother to stop grazing. The alphas, for their part, showed no interest in the bison and moved on.

I later realized there were no bison where these wolves came from. In Alberta the Crystal wolves would have specialized in hunting elk and deer. When young pups start to accompany adults on hunts, they learn what prey animals to pursue by watching older pack members. In the alphas' home country, dinner looked like deer and elk. Biologists call that the predator's prey search image. As wolves travel, they search for animals that match what they think suitable prey looks like. For the Crystal alphas, the Yellowstone bison did not yet fit that image.

Unlike their parents, the four yearlings were curious about this new species, and they began following a big bull that was walking toward the herd. Soon the lead black yearling got to within fifteen yards of him. The huge bison stopped and looked back at the approaching wolves. All four yearlings paused, then three of them, including the little gray, continued forward. The lone bull moved on and soon joined the other bison. As the three yearlings trotted toward the herd, several bison looked up. The brothers stopped and milled around, hesitant to get any closer now that many of the bison were staring at them. The alpha pair were intently watching the yearlings and the herd. At that moment, the bison charged. Clearly intimidated, all four young wolves spun around and ran back toward their parents.

The pack regrouped and continued on its way. The alphas soon spotted a herd of about 150 elk, a perfect match for their prey search image. A cow elk might weigh up to 500 pounds and a big bull as much as 700 pounds. That is much bigger than a wolf, but a more reasonable size for a pack to target than a bison. When the elk ran off, the wolves did not give chase. Instead they advanced slowly. The elk stopped and looked back at the pack, then moved toward the wolves. Since wolves had been back in the park for only about six weeks, these elk were probably still trying to figure out how dangerous they were and how to react to them.

The herd got to within fifty yards of the pack. At that point the elk must have decided that the wolves were a threat, and they began to run again. That set off two of the black yearlings. They chased the herd, but only at a third of their top speed. The rest of the pack stayed where they were and watched. The herd split in two, and now just one yearling was left in the chase. When the elk stopped, he stopped. He stared at the subgroup he had decided to follow, and they stared back at him. I looked back at the alphas. They were still just watching. I got the impression they were evaluating the condition of the elk and had not seen any slow individuals or any that showed signs of weakness that might enable the pack to catch and kill one.

As I accumulated more sightings of wolves interacting with elk, I saw that an average healthy elk can easily outrun pursuing wolves. The top sprinting speed of an elk is around forty-five miles an hour while the maximum for a wolf is about thirty-five miles an hour. To put that in perspective, Olympic champion Usain Bolt runs a 100-meter race at an average of twenty-three miles an hour. In a sprint with a wolf and elk, he would come in last. An experienced older wolf does not waste energy chasing elk that are in good condition because the odds of killing one are too low. The alphas moved on and the yearlings rejoined the pack. I soon lost the wolves in a forest.

I learned a lesson that morning. Sunrise at that time of year was around 5:45 a.m., but there was enough light to see wolves by 5:15 a.m. I had arrived at the parking lot at 6:00 a.m. and had missed forty-five minutes of potential wolf sightings. From then on, I vowed to get up earlier, about 4:00 a.m., so I would have time to eat, get ready, then drive the fifteen minutes from Tower to Lamar to arrive by first light. I did not want to miss anything due to sleeping in.

I did not see wolves the next three mornings, but on the evening of May 16, I spotted a grizzly and a bald eagle as I was scanning for the pack. A herd of elk were staring at a spot in a meadow with concern. I swung my scope in that direction and watched that site. A black yearling, who had been hidden to me, eventually got up there. That sighting taught me to pay attention to prey animals when they are all looking in the same direction.

That evening I saw three species that were on the endangered species list at that time: grizzly bear, bald eagle, and wolf. Neither the bear nor the eagle paid any attention to the wolf, but I would later see both species greatly benefit by having wolves back in the park. Both are scavengers, and we eventually realized that the increase in Yellowstone grizzlies in the coming years was due partly to the free meat they got from wolf kills.

During those early weeks I frequently stopped in at the park headquarters in Mammoth Hot Springs, where Mike Phillips and Doug Smith had their offices. I got to know them and filled them in on the wolf sightings I was having in Lamar Valley. Back then the official name of the reintroduction was the Wolf Restoration Project, but soon everyone was shortening it to the simpler Wolf Project, and that name continues to be used today.

WHILE I WAS having those early sightings of the Crystal Creek wolves, there was a major ongoing story involving the Rose Creek pack. The Rose trio spent their first week of freedom exploring the region near their release point, about five miles east of the Crystal Creek pen. Then the female yearling, wolf 7, broke off from the adults. On her own, she learned to be a master hunter and killed elk by herself. The following year she would join up with wolf 2, one of the black Crystal

Creek brothers, to form the Leopold pack. I was to spend many hours observing the new pair.

After the young female left them, the Rose Creek pair moved east, then northeast, ending up out of the park near the town of Red Lodge, Montana, fifty-five miles from the Rose Creek pen. The alpha female, wolf 9, was nearing the end of her pregnancy and began to restrict her movements in that area. On April 24, 1995, Mike Phillips did a tracking flight and saw the pair together, just inside Custer National Forest. The alpha male left the den site later that day and went out on a hunt.

Doug Smith did a flight two days later and got the female's signal from the same area. The male's signal was not there, so Doug circled the surrounding area. When he finally picked it up, it was on mortality mode, indicating 10 was likely dead. Wolf radio collars have a motion sensor. If no movement is detected over a period of four hours, the beeps per minute double. His body was later found, and Chad McKittrick of Red Lodge was eventually convicted of killing an animal protected under the Endangered Species Act and spent time in jail. McKittrick shot wolf 10 on April 24.

On the day of her mate's death, 9 gave birth to a litter of pups on private land, five miles from where the alpha male had died. Joe Fontaine of the US Fish and Wildlife Service discovered the den site ten days later and confirmed the presence of pups. He got a count of seven. The site was just a shallow depression under a tree. Carcasses were put out in the area to help the new mother survive. Newborn wolf pups cannot regulate their own temperature and must snuggle up to their mother to keep warm. If she became desperate for food and went out on a hunt, her pups might die from hypothermia before she got back to the den.

Since the den site was just four miles from downtown Red Lodge, Mike and Doug decided to recapture the mother and pups and put them back in the Rose Creek pen. On May 18, Carter Niemeyer of the US Fish and Wildlife Service caught 9 near the den in a padded leghold trap placed near some scat from her mate that he had collected from the Rose Creek pen. The crew then went to get the pups.

From his tracking flights, Doug knew that 9 had moved her litter to a new site. Joe walked uphill to that new location, then made low calls, hoping the pups might think their mother was approaching. He heard whimpering in response. He looked in the direction of the sound and spotted a group of pups. They all ran off, except for one that stood its ground and stared at him before following the others into a chamber embedded in a jumble of talus rocks.

Doug, with his thin frame and long arms, reached in and pulled out the three-week-old pups, one by one. There were seven. Seven was the original pup count, but on a hunch that there might be one more, Doug grabbed a stick, poked around, and made contact with something that felt soft. He pulled out the stick and saw a piece of fur caught on the tip, indicating one last pup might be at the back of the den. It was too far to grab by hand, so he got a pair of Leatherman pliers and reached in as far as he could. The pliers closed around something. Whatever it was, the animal struggled against Doug as he pulled it out. It was an eighth pup, a black male.

Mark Johnson, the project's veterinarian, had examined the first seven pups (four females and three males) and determined they were all healthy. That eighth pup, the one that had struggled against Doug, was also in good shape. Mark, due to his years in veterinary practice and working with wild wolves, is very experienced when it comes to recognizing dogs and wolves as they get older. He later told me he believed that the eighth pup grew up to be wolf 21, the most famous male wolf in Yellowstone's history. As an adult, 21 would weigh up to an estimated 130 pounds, but on that day, at an age of twenty-four days, he was just 5 pounds.

The mother and her eight pups were loaded into a helicopter and flown back to the Rose Creek pen. During the flight the pups were free to roam around the interior of the helicopter, but 9 was in a cage. I was out that day and saw the helicopter flying to the site. The wolves were scheduled to stay in the pen for six months, through mid-October, to allow the pups to grow larger and have a better chance of survival on release. Carcasses would be dropped off in the pen twice a week.

The birth of nine pups (eight to the Rose Creek pack and one to the Soda Butte wolves) that first spring was unexpected. No one with the project had thought the wolves would breed while in captivity. But the death of the Rose Creek alpha male so soon after release canceled out much of the excitement over the birth of the pups. However, wolf 10 had made a major contribution to the Yellowstone gene pool before losing his life, and he would live on through the pups he fathered and through the many wolves descended from them. He was the founding father of a dynasty that continues in Yellowstone today.

4

The Little Wolf and the Big Grizzly Bear

O N MAY 18, 1995, the same day the helicopter brought the Rose Creek wolves back to their acclimation pen, I saw one of the three Crystal Creek black yearlings on a fresh elk carcass. Then I spotted the other five pack members on another new carcass. The small gray yearling went to one of his brothers as the black walked off with a piece of meat in his mouth, and the two had a playful wrestling match. 8 snatched the meat away from his brother and ran off with it. He stopped, put it down, and played with it as the black watched. The wolves were so full that day it did not matter who ended up with that piece of meat. There was still plenty left on the carcasses, more than enough for all of them.

One of the law-enforcement rangers who had patrolled the Crystal Creek pen site the previous winter told me the three black yearlings had mercilessly picked on their smaller sibling throughout their captivity. She said they would chase 8, tackle and pin him, then nip at him for a long time. Since there was not much else for the yearlings to do during their confinement, harassing their gray brother was one of the three blacks' favorite pastimes. Usually the bullied wolf would bed down away from his brothers, but they would creep up and pounce on him as he slept. 8 would either run off without fighting back or stand up to them for a moment, then run away.

Since he was the smallest wolf in the pen, the rangers patrolling the area called him "the little guy." The ranger also told me the gray was normally the last to eat when new meat was brought into the pen, a sign of his lowly status. As she told me those stories about 8's hard times in the pen when he was a pup, I recalled the famous quote from philosopher Friedrich Nietzsche, "That which does not kill us, makes us stronger." Would getting picked on and beaten up enable him to cope better with adversity and challenges now that he was growing up?

Knowing the hard times 8 had had in the pen for the ten weeks of the pack's captivity, I was glad to see that his life was becoming more normal. His three brothers had plenty of things to do now that they were free roaming and less time to pick on him.

Later that day, when one of the black yearlings was at one of the new carcasses, a grizzly mother and her two yearling cubs approached. A cub charged the wolf four times. Each time, the black ran off just a few steps, correctly guessing the cub was only bluffing. After a while, the wolf walked toward the carcass, ending up in the middle of the bear family. The other cub charged a few feet at the wolf, then went back to the carcass and fed. That time the black did not even bother running off. I was letting a group of Wyoming schoolkids watch this interaction through my scope, and one boy yelled out to his friend, "This is the most exciting thing I've ever seen in my life." They were from a town known to be anti-wolf, and I was glad to see that watching these wolves was changing the way they saw the world.

FOR THE NEXT few weeks, the six Crystal Creek wolves were visible most mornings and evenings. Instead of doing roving interpretation in far-off places like Old Faithful, I could now drive the few miles from Tower to Lamar, find the wolves, show them to visitors, and tell the story of the reintroduction. News of the wolves' visibility spread through word of mouth and newspaper stories, and more and more people came to the valley to look for them. Soon it was normal to have crowds of two hundred by the side of the road. When the Crystal wolves came into view, people reacted like fans following a popular rock band. Some of them cried when they saw wolves through my scope, and one woman ran to me, as the nearest government official, and hugged me because she was so happy that wolves had been brought back to the park.

I had been very involved with wildlife photography for my fifteen summers in Denali and through my first few years in Yellowstone. I had tried to take telephoto pictures of the wolves after their reintroduction to the park, but found photography got in the way of studying wolf behavior and helping people see the wolves. I was also growing increasingly uneasy with the common practice of trying to get closer to wild animals for pictures. Eventually, I chose to leave my camera equipment at home and give my total concentration to watching the wolves and inviting visitors to see them through my scope.

An unofficial code of behavior developed among the regular wolf watchers in Lamar. People looked for the wolves from the side of the road and did not approach them. Visitors did not howl at the wolves, which is illegal in the park. Instead, they quietly watched them through their binoculars or spotting scopes. Thanks to that respectful attitude, the wolves continued with their normal behavior and often stayed in sight for hours at a time. Wolf watchers with scopes invited other people to look at the wolves through their equipment. That sense of respect and sharing created a very positive experience for everyone. I had never seen anything like it in the other parks I worked in during my then twenty-one years with the National Park Service.

The experience of viewing wolves also attracted people from many different levels of society: working class, middle class, billionaires, and movie stars. One morning when the wolves were in view, a van pulled into my lot. I walked over and told the people they could join me and see wolves through my scope. A tall, take-charge man rushed up, saw the wolves, then asked if his wife could look. When she had seen the wolves as well, the man thanked me and introduced himself. He was Ted Turner and the woman was Jane Fonda. She later wrote me a very gracious thank-you letter.

As I had more sightings of the pack, I concentrated on getting to know the wolves as individuals. I was particularly interested in the four yearlings, and I soon learned that they loved to play. One evening when I was watching the three young blacks near a fresh carcass, I saw one of them go up to his brother and do a play bow in front of him. That seemed to be an invitation to play a game of catch me if you can. It worked, for the second black chased his brother. Later one of them picked up an old elk antler. Another yearling came over and stole it from him, but he soon ran back and the two played tug of war with it. When the alphas moved off, the three yearlings followed, still playing. One would turn around and prance in front of the one behind him, which would set off another playful chase. Both then took turns with first one, then the other in hot pursuit.

A few days later I found two of the black yearlings at another carcass. One grabbed a piece of meat, tossed it in the air, then leaped up and caught it in his jaws. He dropped it and pounced on it like it was alive and trying to get away. After that he ran off with the meat, threw it up again, and caught it on the run. I later began a list of games young wolves play and called this the tossing game.

Another black yearling ran in and chased the first one. The lead brother dropped the meat and the other black grabbed it. When he ran off with it, the first yearling chased him. They reversed the chase and the second black pursued the first. Then the lead black suddenly stopped and lay down in tall grass. As the other wolf ran in, the one hiding in the grass jumped up and wrestled him to the ground. I called that the ambush game.

Later, as the two stood side by side, one brother suddenly ran off. That looked like a dare for his brother to chase him. The other black accepted the challenge and went after him at top speed. The two took turns chasing each other in straight lines and in zigzag patterns, a game of catch me if you can. They ran, pranced, and twirled around in front of each other. It did not matter who chased whom, the point of all that playing was not to win, it was to have fun. The best word to describe the behavior of the yearlings was *exuberant.* As I watched them, I had a thought: they loved being wolves.

All that play served a purpose. I later saw a cow elk chase one of the yearlings. Although she could outrun the wolf in a straight-out race, he zigzagged back and forth so nimbly that she soon gave up in frustration. The vigorous play chasing had prepared him to outmaneuver the cow. At times the yearlings actively invited elk to chase them. I saw them do play bows in front of elk to get them to initiate a pursuit, then easily get away using the tricks they had perfected during their play sessions. It was like they were showing off.

As I watched the yearlings play that spring, I thought of how Yellowstone had become a promised land for the Crystal Creek wolves. No human in their new territory was trying to shoot, trap, or snare them. All they had to do was live the lives of wild wolves.

ONE MORNING I found 8 walking through Lamar Valley by himself. Five cow elk spotted him and chased him. As he ran off, he nervously looked back over his shoulder, saw they were gaining on him, and tried to run faster. They sped up as well. The cows got to within a few yards of him, then lost interest and veered off. A bit later he saw a big bison bull bedded down in a meadow. He dropped into a low crouch and stealthily approached the bull from behind. Soon he was within a few feet of the bull's rear end, but seemed unsure of what to do next. The 2,000-pound bull casually turned his head and glanced at the insignificant small wolf behind him. Unimpressed, he turned his head back and resumed chewing his cud. Now the gray yearling was even more uncertain what he should do. At that moment, the bull flicked his tail to drive off some mosquitoes. The wolf spun around and ran off. If 8 was trying to figure out if this animal might be prey, he evidently decided that this bull was way too big for him to challenge.

Although he did not impress me that morning, I saw a different side of the young gray wolf that evening. He and two of his black brothers were playing and chasing each other. All three suddenly stopped, looked west, and ran into a stand of conifers in that direction. I got glimpses of animals running back and forth in the trees for a few moments, before I briefly lost sight of them. Suddenly, one of the blacks came running out of the forest with an elk calf carcass in his mouth. The other black, then the little gray, appeared a moment later, running in the same direction. Then I saw the grizzly. It was right behind 8 and gaining. The bear was huge compared to the wolf. It looked like a dinosaur chasing a kid in a *Jurassic Park* movie.

Apparently, the bear had killed the calf in those trees. The first yearling must have grabbed its prize while the other wolves were distracting the bear, which was now closing in on the small gray. I visualized the bear swatting the wolf with a front paw, knocking him down, and killing him. Anticipating what was going to happen, I tensed up. Based on what

I knew of 8 and his history of being bullied by his bigger brothers, what happened next took me totally by surprise. I saw him stop, turn around, and confront the grizzly. Startled by this move, the bear pulled up abruptly. The two animals stood there, a few feet apart. It was like watching David standing up to Goliath. The bear looked like it couldn't figure out what to do next as the wolf glared at him in defiance.

While the unlikely hero was confronting the bear, the black yearling with the calf was making good his escape with the other black right behind him. Both disappeared into a thick forest. I looked back at the little gray and the bear. They were still staring at each other from close quarters. Then the wolf turned around and casually trotted away. He seemed totally confident that the grizzly would not renew the chase.

The bear sniffed the ground and the air. Unable to figure out where the wolves had gone with its kill, it wandered off in the opposite direction. I later saw the three yearlings come back out of the forest. The yearling with the calf bedded down and fed on it while the other black and 8 lay down near him, respecting their brother's right of possession.

That episode showed me there was more to 8 than I had first thought. He was the smallest yearling, the one the bigger brothers had picked on, but he was also the one who had had the nerve to stand up to a huge grizzly and get away with it. I realized that none of the other Crystal wolves, not his brothers or his parents, had seen him turn around and confront that bear. I was the only witness to his courageous behavior. Years later I heard Dwayne "The Rock" Johnson say something that applied to the little wolf that day: "Being a hero means doing the right thing, even if no one is watching you."

A few days later I saw 8 lead the pack on a chase of a cow moose, another indication of his rapidly developing maturity.

On July 5, I went out early to Lamar Valley and spotted 8 with two of his brothers. They were wrestling each other in different combinations, and the gray held his own. Later one of the blacks tripped, tumbled, and rolled several times as he was chasing his smaller brother. Seeing the black on the ground, 8 ran back and playfully pounced on him. The two sparred with their jaws until the black managed to squirm out from under him. The gray chased him for a while, then led both his brothers off, and I soon lost them in a forest.

That was my last sighting of the Crystal Creek wolves for the next few months. The elk left the valley to feed at higher elevations, and the wolves followed their migration. The tracking flights Mike and Doug did during those weeks found the pack roaming far and wide. They were often spotted twenty miles to the south in Pelican Valley, just north of Lake Yellowstone. I wondered what would become of 8. He was the lowest-ranking male in his family, but was exhibiting qualities that might make him a successful alpha male of his own pack if he found a mate and a vacant territory. I also thought about his three brothers. The coming year would be a critical one for the four yearlings and likely reveal their long-term fate.

5

The Rose Creek and Crystal Creek Pens

THE PLAN FOR the Rose Creek alpha female, 9, and her pups to stay in the acclimation pen through the fall of 1995 was threatened when a late July windstorm blew down several large trees just outside the enclosure. Two landed on the fence, creating a pair of holes. The damage was not discovered until a few days later, when Doug rode in on horseback carrying elk meat to feed the wolves. By that time, all eight pups had gone out through the openings. Luckily their mother had stayed inside the pen, and since the pups wanted to be near her, they were all still in the area. Mike and other personnel joined Doug and tried to recapture the pups.

At first, they could not see any of them. Mike decided to lure them out by howling. His plan worked and the pups ran out from the nearby trees, thinking the howling was from their mother. Three of the pups went back through the holes in the fence. After the crew closed the openings, they tried to capture the other five pups. They caught two and put them back in the pen. The last three got away but stayed near the pen from then on. Park Service crews left meat outside the pen for those three pups each time they brought carcass parts into the repaired pen.

On October 9, Mike and Doug went up to put radio collars on the five pups in the pen. When they got there, they saw that six pups were now inside. The only way that sixth pup could have gotten in was to climb the ten-foot-high fence surrounding the enclosure, then jump down. They captured the six pups in large fishing nets and placed radio collars around their necks. The average weight of the five-and-a-half-month-old pups was 65 pounds.

In those early years of the Wolf Project, all young pups and uncollared older wolves were assigned numbers. Some were later radio collared, but most were not. That system became impractical when we lost track of many of the uncollared wolves, due to death or dispersal. Eventually the system was changed so that only collared animals were given numbers. As wolves settled in sections of Wyoming and Montana adjacent to the park, we shared our numbering system with the Wyoming Game and Fish Department and the Montana Department of Fish, Wildlife, and Parks. If they planned a new round of wolf collaring, they would contact our office, find out the last number we had used, and assign the next consecutive numbers to the wolves they captured. The collars allowed us to get a signal from a wolf from as far away as ten miles if it was on a high ridge. If the wolf was behind a ridge, the signal would be blocked and probably would not be picked up even as close as half a mile away.

In September, I twice helped Mark Johnson, the project veterinarian, when he went into the pen to feed the wolves. We carried meat up to the pen, opened the gate, dropped off the pieces, then left as quickly as we could to reduce the chances the wolves would get used to our presence or associate people with the sudden appearance of food. When we entered the pen, the mother and pups ran to the far end of the enclosure, then raced back and forth, trying to get farther away from us. After we left, they soon calmed down. As they walked around the pen, they discovered the meat and fed on it as they would on finding a carcass in the wild.

During my moments in the pen, I briefly glanced at the mother wolf and pups, then concentrated on getting out. The first time I was in the pen I saw a big black wolf I assumed was the adult female. Then I saw a bigger black that was obviously the mother and realized my mistake. The other big black wolf was a really large pup. After Mark and I left the pen, we looked around for the two pups still at large but did not find them. When we went back to feed the wolves a week later, I spotted several large wolf droppings outside the pen that looked old. They were probably from alpha male 10 when he had waited patiently for the two females to leave the pen and join him.

Mark fed the wolves far more often than I did. Years later he told me a story that profoundly affected me. He had gone into the pen to leave meat for the pack. After dropping off the meat, he noticed that one of the black pups was acting differently from the other wolves, who were running around at the far end of the pen. That pup positioned himself halfway between Mark and the rest of the family, then repeatedly circled around him. To Mark, it looked like the young pup was acting like the pack's alpha male, protecting his mother and siblings from a threat. The pup never approached him, and Mark did not feel threatened, but the message was clear: do not come closer.

Mark realized that he had seen that behavior before. When the original three wolves had been in the pen the previous winter, the big male would get between Mark and the two females, then circle around him in a calm and confident manner. That black pup had never known his father but was behaving the same way the alpha male had to protect his family. The pup was literally walking in his father's footsteps, doing what his father would have done if he had still been alive. As I mentioned earlier, Mark is an expert at identifying dogs and wolves as they get older. He finished his story by telling me he believed the brave pup who took on the responsibility of defending his pack was 21, who would grow up to be the park's heavyweight champion. I then realized that the big black pup I had seen when I had been in the pen was also 21.

THE STORY OF the Rose Creek wolves, the killing of wolf 10, and the return of the mother wolf and her eight pups to the pen became well known to the public through many reports in the media. In late August, President Clinton and his family were vacationing in Jackson, Wyoming. It was his administration that had approved Yellowstone's wolf reintroduction proposal. White House staff had contacted the park superintendent and asked if the Clintons could come to Lamar Valley and see the Rose Creek wolves. On August 25, I drove by the Yellowstone Institute and saw the presidential helicopters parked nearby. Mike and Doug took the first family up to the pen, and the Clintons helped bring in meat for the other famous family: wolf 9 and her eight pups.

Because of all the media coverage, there was tremendous public interest in the wolf acclimation pens. In addition to helping visitors see wolves in Lamar Valley, doing roving interpretation there, and giving my evening slide shows in park campgrounds, I also led twice-weekly hikes to the Crystal Creek acclimation pen. Normally, ten to thirty people show up for park ranger–led hikes. I had up to 165 visitors join me on the pen walk. I had done so many nature walks during my Park Service career that I was confident managing the large crowds that came with me. On the way to the pen, I paused periodically and told the wolf reintroduction story in stages. I always stopped at an old aspen grove and had people look at the hundreds of shoots coming up from the roots. Aspens usually regenerate from roots of existing trees rather than from seeds, and I pointed out how every shoot had been browsed to death by hungry elk in the winter months.

We now know that after the last Yellowstone wolves were killed off in 1926, the elk population skyrocketed, because one of their main controlling factors had been eliminated. (Cougars also prey on elk, and the rangers killed them off during that time as well.) In the early 1960s, the Park Service brought in range management experts to analyze the vegetation in the northern section of the park, an area known as the Northern Range. In their 1963 report, they estimated that the carrying capacity for elk wintering in that region was about five thousand, well below the number living there at the time.

Going all the way back to the 1920s, the Park Service live-trapped and shipped out Yellowstone elk to any state or Canadian province or zoo that wanted them. Rangers also shot elk to reduce the overpopulation. Due to the controversy over shooting elk, that program ended in 1968. The capture program was also shut down. By that time, 26,400 elk had been removed from the park, either dead or alive. The elk population rapidly increased from that point on. Just prior to the arrival of the wolves from Canada in 1995, there were 19,000 elk wintering in the Northern Range, nearly four times the estimated carrying capacity of the area. The overpopulation led to extreme overbrowsing of aspen shoots, extensive damage to willows growing along creeks and rivers, and erosion along waterways due to loss of vegetation.

As we approached the pen area, I finished the story of how the wolves had been placed in the acclimation pens and later released into the wild. Then I said we would go around a nearby rocky knoll to a point where they could see the pen. I added that I would not talk when we got to that viewpoint, because I wanted each person to have a quiet moment to see the pen and think of its significance. We silently walked around that knoll, and the hikers finally saw the pen. After hearing so much about it, seeing the pen was an emotional event for visitors, especially when they noticed the panel where the exit hole had been cut. That was the exact spot where the first pack of wolves had come out of the pen to become permanent Yellowstone residents. It was the wolf equivalent of Plymouth Rock.

Due partly to the return of wolves, and partly to other factors such as rising numbers of mountain lions and bears, increased human hunting north of the nearby park border, competition from larger numbers of bison, and climatic changes, winter elk numbers in the northern section of the park dropped in the coming years. They eventually stabilized in the six thousand to seven thousand range, a level more sustainable for the ecosystem.

As the wolf packs became more settled in the park, I continued to lead hikes up to the pen site and often hiked up there on my own. Within a few years the aspen trees near the pen site were producing tens of thousands of surviving shoots each spring, which soon formed a forest nearly as dense as a bamboo thicket. Willows also began to flourish along Crystal Creek, and beaver, which need aspen trees and willows for food and building materials, moved in and colonized the area. When documentary filmmakers come to Yellowstone to do stories on the wolf reintroduction, Doug Smith takes them to that creek to point out the amazing recovery of the ecosystem.

My Park Service job ended in early September, but I stayed on in the park to look for wolves. That fall I got together a group of volunteers, and we helped Doug carry fence panels to a new acclimation pen site in the Blacktail Plateau area, about ten miles west of Tower. Mike and Doug planned on bringing in four more packs from Canada, and two new pens needed to be constructed. The Rose Creek and Crystal Creek pens would be reused for the other two packs.

6

The Rose Creek Wolves Get a New Alpha Male

I HAD ONE LAST sighting of the Crystal Creek pack before I left the park for the winter. On October 5, 1995, I saw five of the six Crystal wolves traveling through Lamar Valley. It was the forty-fifth time I had seen the pack that year. The missing member was the gray yearling, wolf 8. Recent tracking flights had often spotted him apart from his family, exploring new country. In February, he would be old enough to breed a female and father pups. I wondered if he was looking for a mate.

A few days after that sighting, I started getting ready to do a lecture tour in Japan on wolves. Gray wolves were once native to Japan, but they had all been killed off by the late 1800s. A wildlife professor, Dr. Naoki Maruyama, who had started a campaign to bring back wolves, had asked me to help by giving talks on the success of our wolf reintroduction program.

I stopped in at the Wolf Project office one last time before leaving, and Doug told me an extraordinary story involving wolf 8. Ray Paunovich was making a documentary on the wolf reintroduction. On the morning of October 11, he had been up by the pen and had seen the two pups still outside the fence begging food from 8 and playfully romping around him. He was the first adult male wolf the pups had seen, and they seemed to be fascinated with him. When I interviewed Ray for this book he told me: "8 acted very friendly with the pups. My impression was he had already made friends with them. They were hanging out together and were very comfortable with each other."

Right after that encounter, Doug, Mike, and a few others went to the pen and opened the gate so the mother and the six pups inside could go free. By late morning, 9 and all eight of her pups were together, and there was a tenth wolf in the group: 8. After that, he was seen traveling full time with the Rose Creek alpha female and her pups. He was now functioning as the pack's alpha male, a place in life he might never have achieved if not for the benevolent way he had treated those first two pups.

Over the years, I have tried to envision what had happened. I visualized 8, after leaving his family, heading up Rose Creek to investigate the wolf howling he heard from that area. On getting there, he would have seen the two free-roaming pups. He had always been the smallest wolf in his world, but now, for the first time, he saw wolves that were smaller than he was. That sighting probably triggered a paternal instinct and he befriended those pups. Their mother, who was still in the pen, would have watched when he interacted with the first two pups. She would then have been welcoming to him when she and the other six pups were released. Due to his youth and size, 8 was not the most ideal candidate to be her new alpha male, but he had been friendly and playful with her pups, so the mother wolf accepted him into her family. She was looking for someone with a heart of gold, and she found that in him.

That day 8 and 9 started to form a long-term pair bond, something only 3 to 5 percent of the roughly five thousand mammal species in the world do, something humans and wolves have in common. He was still a yearling, about sixteen years old in human terms, when he joined the pack. Over the years I have watched many yearling wolves at dens and recorded endless hours of them playing with pups. It is obvious that yearlings love to interact with them. Given my observations, I am sure 8 behaved the same way with 9's pups and was much more playful with them than an older male would have been. It was probably that very thing that impressed her.

I later read how the act of nursing, in both humans and animals, releases the hormone oxytocin in mothers and in their offspring. It also is released when a mother cuddles or strokes her baby. Sometimes called the love hormone, it strengthens the bond between them. One study even described how oxytocin caused a mother to have an irresistible desire to interact with her baby. Oxytocin is also released in fathers, and in their sons and daughters, when they play together, especially when a father and son engage in roughhouse play. For both sexes, higher oxytocin levels correlate with increased empathy, attachment, and altruism. Every time 8 played with the Rose Creek pups, the hormone oxytocin enhanced the emotional bond between them, particularly when he engaged in roughhouse play with his four adopted sons.

When 9 invited 8 into her pack as the new alpha male, one of his primary responsibilities was to be her personal champion, like a medieval lady of the court asking a knight to be always ready to defend her. As the Rose Creek alpha male, he would have to stand between his new family and any threat and defeat it. Most people who knew 8 at that time would probably have said that the alpha female had made a questionable choice. One of his bigger brothers would have been a far better candidate. But I had seen him stand up to that grizzly and felt differently.

As I thought about 8, I was reminded of something said to the hobbit Frodo in J. R. R. Tolkien's *Lord of the Rings* trilogy: "Even the smallest person can change the course of the future." 8's physical size was not his defining feature. What mattered was the size of his heart. He had the big heart of an athlete who may have modest skills, but never gives up.

That was the first time we documented a case of an unrelated male joining an existing pack after the death of its alpha male and adopting and raising the pups as if they were his own. For most predator species, the normal behavior for a new male would be to kill the young of the prior male, breed the female, then help raise the young he sired. That is the custom in African lion prides, for instance, but male wolves are different. In all the cases I later witnessed, new alpha males helped raise the pups born to the previous male.

That behavior was likely a key reason for the successful domestication of wolves by early humans. Nearly everyone knows of a family with a big male dog who is gentle with toddlers and young children, even when they tug on the dog's ears and tail. That tolerance, along with dogs' desire to play with kids and protect them, comes directly from their wolf ancestors.

While writing this section, I was struck by the sequence of events when 8 first came across the Rose Creek wolves. He initially met and befriended the two pups outside of the pen and only later met their mother, who was to become his mate. That meant that his bonding to this new family began with those two pups, not the adult female, an important distinction. When 9 and the remaining six pups came out of the pen, 8 went through the same bonding process with them. For all those pups, 8 would be the only father figure they would ever know. When I later watched the family interacting together, I came to think the emotional attachment and devotion the pups had to 8 would have been no different from the relationship they would have had with 10 had he lived.

ONE WEEK AFTER 8 joined the Rose Creek pack as its alpha male, Mike Phillips and Bob Landis witnessed an event that forced 8 to make a critical decision. Mike saw the five remaining Crystal Creek wolves (the alpha pair and 8's three black brothers) at the eastern end of Lamar Valley. They were trotting west at a fast pace with their alpha male in the lead.

Mike was getting signals from the Rose Creek wolves to the west, toward Jasper Bench. Doug Smith had flown the day before and had seen the Rose Creek wolves on a bison carcass in that area. The Crystal wolves continued trotting west. Then the three black yearlings broke into a run, probably because they had picked up the scent of the carcass. The alpha pair, however, immediately turned around and fled back to the east. They must have realized that they were approaching the larger Rose Creek pack.

Meanwhile, when the black yearlings spotted their gray sibling, they ran to him and had a friendly reunion. The eight Rose pups then joined the four brothers and all twelve wolves enthusiastically greeted each other, wagging their tails and licking each other's faces. The group began to move off toward the Crystal alphas. Mike spotted 9 downhill from her family and what to her were three strangers. As she watched the unknown wolves interacting with her pups, she barked in alarm. When they heard her, the group stopped, and one of the black yearlings started to approach her. 8 followed his brother. The pups stayed behind, probably reacting to the warning calls from their mother.

At that moment, 9 ran up to the black yearling and attacked him. 8 did not hesitate. He jumped in and fought his brother to support his new mate. The Rose Creek alphas were now on either side of the black and both were biting him. The pups stayed a short distance away and did not get involved. When the black ran off, 8 and 9 chased him. All three ran at top speed. After about four hundred yards, the alpha female gave up the chase, but 8 continued to pursue his brother. He gained on the black, before letting him go and returning to his new mate. The alpha pair then trotted to the pups and the family had a big, happy reunion.

Bob filmed that interaction. His footage shows the pups licking 8's face and 9 putting her paws affectionately around his head as she sits up in front of him. She looks like one of the pups greeting the pack's alpha male after he has vanquished an enemy. I thought about how 8 had reacted when he saw her charge at his brother only a few minutes after he had that friendly reunion with him. 8 was now her partner and had to take her side. He turned on his brother and helped her drive him away from her pups. Without question, 8 was now a Rose Creek wolf.

I wanted to stay on in Yellowstone to follow what was happening with 8 and his new family, but there were no winter jobs for me, and I had committed to the lecture tour of Japan. I left the park on October 15. I landed in Tokyo and spent a week in the city speaking on wolves and the reintroduction in Yellowstone. Then I traveled around the country for another week, including the island of Hokkaido, and gave additional talks in those areas.

During my time in Japan, a host took me to a wolf temple. In the shogun era, peasant farmers were forbidden to own any weapons that might be used to rebel against the rulers. The native deer species would come to their farms and feed on the crops. Without any weapons, the farmers had a hard time keeping the deer away. As a clever solution, temples were built throughout the countryside and dedicated to wolves. Farmers would walk to the nearest temple, leave a symbolic food offering to the wolves, then pray that they would come to their farm and kill the deer.

On the way back to the States, I stopped in Hawaii and gave talks on wolves at Hawaii Volcanoes National Park on the Big Island and at Haleakala National Park on Maui. When I returned to Big Bend, I went through my field notes from the summer and found that I had had 138 wolf sightings. I counted each wolf as a sighting, so if I saw all six Crystal Creek wolves, that would be six wolf sightings. During the long summer days in Yellowstone, I would be out looking for wolves by 5:00 a.m. When they were out of sight or inactive in the middle of the day, I would work, rest, or do other things, then come back out in the evening. If I found the pack again, that would count as another six wolf sightings.

I kept track of how long the wolves were visible during each sighting, and that added up to thirty-nine and a half hours. Doug Smith did wolf research for nine summers and two winters at Isle Royale National Park in Lake Superior. In a typical summer season, Doug would hike about five hundred miles through the park's forests and marshes. He said it was a big accomplishment to see one wolf per summer and a sighting averaged less that one minute. Doug saw wolves only three times from the ground during his time there.

I spoke to forty thousand park visitors that summer about wolves, both during Park Service programs and while doing roving interpretation, and I had helped many of them to see wolves. The media had taken a great interest in the reintroduction program, and I was able to get the wolves' story out to an even wider audience through over thirty television and newspaper interviews. I also had 255 grizzly sightings in Lamar Valley. 1995 had been a very good year.

7

The Arrival of
the Druids

I WANTED TO RETURN to Yellowstone for a winter visit to see what the park was like at that time of year and hopefully spot wolves. I signed up for a late January class at the Yellowstone Institute in Lamar Valley entitled "How Animals Survive the Winter," taught by Dr. Jim Halfpenny, a local carnivore ecologist and wildlife expert. The dates of the class coincided with the planned arrival of the second batch of wolves from Canada. To ensure enough genetic diversity, this time wolves would be captured in British Columbia, in the Williston Lake area, 750 miles north of Yellowstone. That was an area where biologist John Weaver had found bison remains in scat at a wolf den, so this second set of wolves had experience hunting bison.

THERE WAS ANOTHER major event that began to play out before I got back to Lamar Valley. In December, one of 8's black brothers, wolf 3, had left the Crystal Creek pack and traveled to Paradise Valley, about twenty-five miles north of the park border. A few days later, he was spotted near a pack of captive wolves in that area that included several females coming into heat. On January 11, a local rancher reported a missing sheep. An Animal Damage Control officer found a partially eaten lamb the next day. 8's brother was in the area and likely responsible for killing the lamb. He was captured two days after that and later released in the middle of the park in hopes he would stay there.

The errant wolf returned to the sheep ranch on February 2. That night a sheep was attacked, and the black was tracked to within two hundred yards of the ranch's barn. He stayed near the ranch for the next few days. The original protocol set up in the park's wolf management plan called for giving a problem wolf a second chance after it kills livestock. If the wolf returned and caused additional problems, it would be destroyed. Because he had failed to take advantage of the second chance given him, 8's brother was shot and killed by an Animal Damage Control officer on February 5. The second part of the prophecy had come to pass. One of the four sons of the mighty alpha male had indeed died young and in disgrace.

Another Yellowstone wolf was lost that winter. One of the Rose Creek male pups was hit and killed by a delivery van on December 19 in Lamar Valley. The other seven members of the litter survived their first year.

IN JANUARY 1996, I flew from Texas to Bozeman, Montana, then drove down to the park, where I had a brief sighting south of the Institute of the Crystal Creek alpha pair and one of the black yearlings. The wolves paused to watch some of the thousand elk in the valley, then turned around and disappeared into a forest.

I settled into Jim's class to learn more about how wildlife copes with the harsh winter conditions in the park. While we were in the classroom on January 28, we heard a commotion and walked outside to see what was going on. A big truck pulling a horse trailer had just arrived at the parking lot. It contained the five members of the new Druid Peak pack, named after the 9,583-foot mountain northeast of the Institute. Each wolf was in a sturdy metal cage, four feet long, two feet wide, and three feet high.

We had already heard the story of one wolf in that truck: wolf 38. He was a big male and weighed 115 pounds. Somehow, he had torn his cage apart during an earlier stage of the transport. A crew member checking on the wolves had found him casually strolling around inside the trailer. He had to be tranquilized and put in another cage for the rest of the trip. Everyone involved with the shipping of the wolves was impressed by the big wolf and a bit intimidated. His story reminded me of King Kong tearing off his chains and escaping into New York City. This guy, I concluded, was not a wolf to mess with. If he could tear apart that cage, what could he do if he got in a fight with another male?

The other four wolves in the trailer came from the Besa pack: a white alpha female (39) with three female pups, one gray (40) and two blacks (41 and 42). The alpha male from that group was not captured and probably had been killed by trappers. The formidable new male was from another pack. He would be put in the Rose Creek pen with the four females in hopes that he would bond with the alpha female, just as the lone male in the Rose Creek pack had with 9 the year before. We went back to our class, happy that we had just witnessed the arrival of a new pack.

Three other packs were brought down from British Columbia at that time. The Chief Joseph pack was placed in the Crystal Creek pen. There were four wolves in that group: an adult male, an adult female, and two pups. The pen we had helped prepare last fall in the Blacktail Plateau area would hold two adults, one male and one female. They would eventually be named the Lone Star pack. The fourth pack, known as Nez Perce, was put in a new pen near Madison Junction, about twelve miles north of Old Faithful. There were six wolves in that group: two adults and four pups.

Those seventeen wolves from British Columbia, along with the fourteen that had arrived from Alberta in 1995, added up to a total of thirty-one. The park had authorization to bring in more wolves, but those original ones did well enough that no additional wolves from Canada were needed. I checked the weight of the seven male pups in the two batches of wolves and 8 was still the smallest. Of the eight female pups, only one was smaller than him. Another one outweighed him by 28 pounds.

Wolf 8 and his new mate were seen breeding in late February, and 9 gave birth to three pups in April. 8 was just two years old, equivalent to a man at age twenty, when those pups were born. He was now responsible for protecting and feeding his mate, her seven yearlings, and the three newborn pups he had fathered. That added up to eleven wolves. It would be a big job for a small wolf.

I FINISHED UP my third winter in Big Bend and began my long drive north to Yellowstone. I arrived in the park on May 12, 1996, a few days before my job started. The first thing I did was drive out to Lamar Valley to look for the Crystal Creek wolves where people said they were denning, a few miles east of the Yellowstone Institute. I knew the alpha pair in that pack well, and I was anxious to see how the family was doing. I did not see them or the newly released Druids.

Early the next morning, I returned to Lamar and climbed the steep hill above the confluence of the Lamar River and Soda Butte Creek to look out over the valley. I soon spotted 8's mother, wolf 5, about a half mile away. She raised her head and howled several times. Then she walked off slowly, limping on her front left paw. I spotted one of 8's black brothers out in front of her. Now two years old, he was the only one of the original four brothers who was still in the pack.

The female frequently stopped to howl and look around. I figured she was looking for the pack's alpha male. The young male, wolf 6, was traveling at a normal pace, and the limping female was having trouble keeping up. She bedded down, and I sensed something more than a sore paw was wrong with her. 6 turned around, went back, and sniffed her. She got up and followed him when he trotted off.

The male went to the bank of the Lamar River, saw two Canada geese in the water, jumped in, and dog-paddled toward them. They easily outswam him and got away. Swinging away from the river, the wolves headed toward a small group of cow elk. The male picked one out, chased her, and easily caught up, even though he was running at only half speed. The cow had something wrong with her, and 6 had detected her vulnerability. He ran alongside her for a few moments, then leaped up and bit into the side of her neck. She stopped and stood still as the wolf balanced on his hind legs, maintaining his grip on her throat. A wolf has four sharp canine teeth and 1,500 pounds of pressure in its jaw. A bite to the throat of an elk can kill it in a few minutes.

In what seemed like a gentle move, the wolf twisted his jaw and upper body and forced the elk to the ground. She did not resist. As he maintained his hold, I could see through my scope she was still breathing, but the force of his grip was slowly suffocating her. When he let go, four minutes after beginning his chase of the cow, she was dead.

The young wolf started to tear into the elk's underside, but soon walked off to check on the alpha female, who was watching a cow bison with a newborn calf. When the cow moved toward her, she had to retreat. 6 returned to his kill and the female followed. They both fed, but 5 walked off after only a minute, another indication that she was either injured or sick.

6 pulled a choice part off the carcass and carried it toward the female. She excitedly trotted to him. I lost both in a gully, then saw that she was eating the prized tidbit. The male was standing a few yards away watching her eat. It looked like he had brought her food because she was hurting. He returned to the carcass and fed. Later both wolves walked off to the south. The female did a squat urination next to a tree, and the male did a raised-leg urination over it. That double-scent mark would usually be done by the pack's alpha pair.

The Crystal alpha male, wolf 4, was nowhere in sight, and I now wondered if something had happened to him. I noticed one more thing about the female. She had distended nipples, a sure sign she was nursing pups. Perhaps her mate was at the den while she took a break and went out hunting with the young male. But why was she limping?

Later in the day, I went to the Wolf Project office to tell the staff what I had seen. Doug told me that he had talked with a visitor who had seen a wolf pack chasing a single black wolf near the Crystal den a few days earlier. Black was the color of the alpha male's coat. Then Doug discovered 4's radio collar was transmitting a mortality signal. Doug and a crew hiked out to that area and found his body. They determined he had been killed by other wolves, probably on May 7, and the Druid wolves were the prime suspects. Apparently, the Druids wanted Lamar Valley as their territory, even though the Crystal Creek pack had claimed it a year ago. The Druids must have found the Crystal den site, then attacked and killed the alpha male. The alpha female's injuries had probably been inflicted by the Druids at the same time.

But what had happened to the pups? Doug and other Wolf Project staff searched the likely den area, but never found a den or any pup remains, nor were the pups ever seen with the two surviving adults. But the female's distended nipples and the fact she had been based in that area suggested that she had a den and pups there. During her field work in Glacier National Park, longtime wolf researcher Diane Boyd twice saw a mother wolf bury a dead pup. Might the female have done that if her pups had been killed by the Druids?

Wolves do not normally breed with close relatives, and 5 was understood to be the mother of the young male. What would happen to the pack in the next breeding season? Mother and son might split up to seek out unrelated mates.

All these developments were difficult emotionally for those of us who had gotten to know the Crystal Creek pack so well over the past year. They were our home team. Now the pack was in danger of going out of existence, because the last two members were mother and son. And it was all due to the newly arrived Druid wolves. People began to refer to them as the bad wolves of Lamar. To us, it was as if a band of outlaws had ridden into town and taken over.

When I later had sightings of the five-member Druid pack, I concentrated on watching their big male, the one who had torn apart his metal cage. He had probably killed the Crystal Creek alpha male. As a park ranger, I tried to restrain my natural inclination to dislike 38, but it was hard to be objective. The Crystal Creek wolves had a high-quality territory with large numbers of prey animals. The Druids had outnumbered and outfought them, killed their alpha male, and taken their territory. There was no biological reason to object to that. Like countries waging war against each other over territory, these wolf packs were doing what other wolves had done for thousands of years. As the Wolf Project later documented, aggressive territorial behavior tends to limit the number of wolves in a given region to the area's carrying capacity.

The Crystal Creek wolves soon traveled twenty miles south to Pelican Valley and claimed that area as their new territory. They had discovered that lush valley in the summer of 1995 and frequently returned to it in the fall and winter. Now the two remaining members of the pack made it their

8

A New Pack
Is Formed

I WENT TO PARK headquarters to check in and found out there was a change of plans. For the summer of 1996, I would be living at Madison Junction, thirty-seven miles south of Mammoth. That was a long way from Lamar Valley. The naturalist division wanted me to give wolf talks at Old Faithful so we could reach larger numbers of visitors, and a slide show every Friday evening at Madison Campground, sixteen miles north of Old Faithful. They also scheduled me to do two wolf-themed hikes each week at Harlequin Lake in the Madison area. The rest of my time was scheduled for roving interpretation with my wolf pelt in the geyser areas.

I contended it would be better if I were based at Tower so I could continue to help people see wolves in Lamar, but my schedule of programs had already been published in the park newspaper, and it was too late to change anything. That was on May 13. I needed to move temporarily to an old trailer at

Old Faithful on May 29, then switch into another trailer at Madison after it was renovated. I had the next fifteen days to look for wolves in Lamar.

On May 16, Mike Phillips took a group of us up to South Butte, on the south side of Blacktail Plateau, ten miles west of Tower. Wolf 7, the female yearling who had struck out on her own after stepping out of the Rose Creek acclimation pen, had settled down in this area in the spring of 1995. The tracking plane had occasionally found her back in Lamar Valley in the fall and early winter. When she appeared there, one of 8's big black brothers, Crystal Creek yearling wolf 2, was often nearby. In January, he followed the lone female back to Blacktail and they paired off, creating the first new pack to be established by the reintroduced wolves. In honor of wildlife biologist Aldo Leopold, who in 1944 had been the first to suggest wolf reintroduction in Yellowstone, the pair was named the Leopold pack. They were now raising three pups in a forested area near South Butte.

Mike pointed out the new wolf acclimation pen south of our position. That was where we had carried the fence panels last fall. Two unrelated wolves, a male and a female, had been placed there in January. Later the Leopold wolves had chosen their den site just a mile away. The original plan had been to release the two wolves in the pen in the area, but the Leopold den meant another site had to be found to set them free. A temporary pen was put up south of Old Faithful, near the Lone Star Geyser. The pair were moved to that pen, soon released, and named the Lone Star pack. Not long after leaving the pen, the female stepped into a hot spring and later died of her injuries. The male left the area, wandered throughout the park, and died two years later well to the east at Lake Yellowstone.

I signed up as a Wolf Project volunteer, which allowed me to use telemetry equipment to help find and monitor the Leopold pack when I was not working my regular naturalist job. As a volunteer, I had to fill out detailed forms on the wolves I observed and list the times of all their behavior. I spent many days on South Butte throughout the summer, and my time there gave me a chance to study parental care of young pups, pup behavior, and wolf hunting methods. In addition, I got to watch a new pair of wolves, together only a few months, to observe how they interacted in the early stages of their relationship and witness how they bonded.

Two days after going on the hike with Mike, I went back up on South Butte and spotted the Leopold alpha female bedded down just north of the small forest where she had her den. That was my first sighting of her, for I had not seen her at all in 1995. She later got up and went to a nearby elk carcass. Fog rolled in and I lost sight of her. When it lifted, 7 was gone and was probably in the den forest attending to her three pups.

I stayed all day and in the early evening spotted the black alpha male. He approached the den forest from the west and disappeared into the trees. Right after that, the female also came in from the west. As she moved toward the forest, she stopped and stared into a gully at the edge of the trees, then ran down into it, like she was excited at seeing her mate.

Later both parents came out in the open. The female lay down. 2 came toward her and she rolled on her back. As he stood over her, she lifted her front paws and gently touched his face. It looked like she was caressing him. He sniffed her belly and probably got the scent of milk from her most recent nursing. She wagged her tail and continued to touch him gently. Then she jumped up and licked his face. When they walked off side by side, 7 pressed her face against his in an affectionate manner. Then she rolled on the ground and tried to get him to play, but this time he walked off and bedded down. He was probably tired from the day's hunting trip.

The female was not done with him. She jumped up, ran the ten yards to him, and lay down beside him. 7 rubbed her jaw on his mouth. He reacted by licking her face. She rolled on her back and put both front paws around his head like she was holding the face of her lover. Later she rubbed her back on his side and licked his face. I watched the affectionate interaction with astonishment and gratitude. Never had I witnessed such an intimate moment between a pair of wolves. I realized how such affection would strengthen the bonding of the pair and act as an incentive for the male to faithfully go out on solo hunts, make kills, and bring back food to his mate and their pups.

Toward the end of the evening, the pair walked over to the carcass and fed. I did not see it that day, but when wolf parents return to the den after feeding, the pups run over and greet them by jumping up and licking the sides of their jaws. That triggers the adults to regurgitate meat they are carrying in their stomachs. The wolf lowers its head and brings up chunks of freshly swallowed meat. Each pup grabs a piece, runs off with it, and gulps it down. If the mother wolf stays back at the den to tend to the pups while the father goes out to hunt, he feeds her the same way. That face licking by wolf pups is the reason pet dogs lick the faces of their human friends when they return home. For dogs, it is a greeting, but the behavior originated with their wolf ancestors and had a different purpose: begging for food.

The following day I saw both parents carry meat from that nearby carcass to the den forest. It is more efficient for adults to eat as much meat as they can and bring it to the pups internally than it is for them to carry big pieces in their mouths, especially if the carcass is a long distance from the den. Big male wolves can quickly swallow as much as 20 pounds of meat if their stomach is empty, then go directly to the mother and pups and share it with them. In this case, however, the carcass was so close to the den that the parents could both swallow a lot of meat and carry an additional portion in their mouths.

On May 21, the alpha pair went out on a hunt. They encountered a lot of elk, chased some of them, but could not catch up with any. That lack of success did not seem to bother them too much because later 7 turned to her mate and bounced up and down on her front paws in front of him like she was dancing. Then she romped off to a nearby snowbank, acting like a puppy wanting to play. The male joined her, and they both rolled on the snow patch, one of many that had not yet melted on that 7,000-foot plateau.

They walked off, and after lagging behind, 2 ran to catch up with her. She saw him approaching and again bounced around in front of him as she wagged her tail. Then I saw her charge forward, rear up, and bounce off his chest. After that 7 did a play bow that conveyed the message: "Chase me!" She ran off, but checked over her shoulder to make sure he was pursuing her. When she saw that he was not following her, she ran back and bumped against his chest again. He responded by putting his front paws around her neck. Then they ran off, side by side, in a romping playful gait. After chasing more elk, they returned to the den forest to check on the pups.

I went back up to the top of South Butte a few days later and saw the alpha pair leaving the den forest. First the two wolves chased a group of adult pronghorn. They can run up to sixty-five miles an hour, nearly twice the top speed of a wolf, so wolves have little chance of catching one. Then the male spotted a herd of seventy elk and went after them instead. He drove them toward his mate, and I wondered if that was a deliberate strategy. The elk went out of sight in a ravine, then reappeared. At that point, the pair joined forces and pursued the elk together. The big herd split in two, then the subgroup the wolves stayed with split again.

The pair concentrated on a group of eight cows. Soon 7 was chasing one group of cows while the male went after a different one. Both groups easily outran the wolves, and each wolf had to give up. I questioned if the wolves had made a mistake in splitting up, thinking it would have been better if they had joined forces and gone after the same elk. But perhaps all that chasing was intended to test various elk to find one that was slow enough to catch and kill. If so, having each wolf test separate elk simultaneously was an effective strategy.

ONE DAY, I spotted a grizzly sniffing around at the edge of the den forest. A black bear had passed by earlier, and the grizzly was probably on its trail. I lost the grizzly when it went into the den forest, and I was concerned for the pups. It turned out there was no need to worry. In the fall, months after the family had left the site, Mike and I hiked out to the den with a crew filming a documentary on the wolf reintroduction. We saw that the female had chosen a site where numerous big trees had fallen over to create a maze-like thicket. She had dug her den under the downed trees, making it very difficult for a grizzly trying to get the pups. I thought about how 7 was a first-time mother and had never chosen a den location before. She did know about grizzlies and may have visualized the possibility of one digging out her den and killing her pups. Once she had that image in her mind, she apparently conceived of a plan to keep her pups safe from bears: dig out a tunnel under a logjam. All that implied that wolves can think in terms of the future and make plans based on their assessment of what might foil a potential enemy.

After I lost sight of the grizzly, I saw the female heading back to the den. A cow elk stood directly in her path and did not back off as she approached. They had a face-to-face confrontation, and I noticed how much bigger the cow was compared with the mother wolf, at least four times larger. The cow charged at the wolf, stopped, and stamped her front hooves. 7 ran off but came right back to confront the cow again. At that point the cow ran away. The wolf did not chase her, but continued to the den forest, likely to check on the pups and nurse them.

That evening I spotted 2 bedded down near the den forest. He soon got up and went out on a hunt. His nose was to the ground as he followed a scent trail. Five nearby cow elk watched him but seemed to realize that they were not the ones he was focused on. Then I spotted another cow with a newborn calf. Its mother ran away when she saw the wolf, and the calf knew by instinct to run after her. 2 now had them in his sights and immediately pursued them, ignoring the many other elk he passed.

The calf was now lagging behind its mother. I estimated it was halfway between her and the wolf. The mother elk looked back, saw the danger to her calf, turned around, and charged. 2 got to the calf a second or two before she arrived, bit it on the back of the neck, and tried to drag it off. It would have been a killing bite if he had had a few more seconds to hold on. But the cow reached him at that moment and struck out with her front hooves. He dodged the strike, dropped the calf, and ran off.

The cow chased the wolf and repeatedly kicked forward with her front hooves. He zigzagged frantically, barely dodging her deadly kicks. Then he circled around behind her and went after the calf again. The cow turned and chased him and drove him past the calf. They went back and forth around the calf several more times. Two other cow elk ran in and helped the mother drive 2 off. He was outnumbered three to one. Each cow was at least 300 pounds heavier than he was.

A few moments later the wolf was back, confronting the mother elk once again. This time, the two other cows stood by the calf, which by now was lying down. The wolf turned away from the mother, ran toward the calf's two protectors, slipped past them, and got a quick bite on the calf before the pair chased him off. That was his second contact with the calf.

After another confrontation with the mother, 2 got past her and bit at the calf for a third time as he ran by it. All three cows were chasing him now, and all he had time for was a quick nip. Four more cows ran in and positioned themselves between the wolf and the calf. With seven cow elk teaming up to defend the calf, the wolf had little chance of getting it.

At that moment the calf got up and ran. The wolf saw his chance and went after it, darting expertly between the cows. I thought of the times the previous summer when I had seen him and his brothers playing catch me if you can and how they dodged each other. That game was the perfect training for him to survive this dangerous situation. He caught up with the calf and bit it on the back, but then had to drop it as the cows ran at him. That was his fourth contact.

The exhausted calf lay down again and the wolf darted in, grabbed it, and tried to drag it off. The calf's weight slowed him down, and the cows easily caught up with him. He dropped it and just barely escaped a trampling by the lead cow. His fifth contact had not finished the job, but the calf now looked injured. Charging through the cows like an NFL running back trying to get through the defensive line to the end zone, 2 reached the calf and bit it for the sixth time.

He managed to drag the calf off three yards before the cows charged in, and he had to drop it. The situation then got much worse for him. More cows were running in. Now there were thirty of them, determined to protect that one calf. All thirty chased the lone wolf.

The calf then made the mistake of getting up and running after the cows. Immediately, the wolf circled around behind the calf. A cow ran in and blocked him. The calf, exhausted and wounded, lay down and 2 charged at it. Cows drove him off. The calf, head held high, calmly watched the action around it.

Taking a break, the wolf lay down in a small stream to cool off and drink some water. To the thirty cows, it looked like he was giving up. They turned around and went toward the calf. 2 stood in the water for a few moments, seemingly defeated, then charged at the herd. That spooked them and they ran off, right past the prone calf. The wolf saw his opportunity. He ran all out, reached the calf, and delivered a killing bite to the back of its neck.

I looked at my watch to see how much time had elapsed during 2's repeated attempts to get that calf, thinking that it might have been thirty minutes. It was only five minutes, but a very intense five minutes. It was his seventh bite that killed the calf.

At any moment, the new alpha male could have been kicked or trampled to death by one or more cows. He had to take great risks to feed his family, and today he had succeeded spectacularly. But this was only one day. Those rapidly growing pups needed a lot of food every day and would for months to come. He would have to risk his life many more times that spring and summer. Later in the evening I saw the devoted father carrying a big piece of meat from his kill into the den forest, where his mate and three pups would be eagerly awaiting his return.

At first no wolves were visible. Then we saw one of the black yearlings walking around the den site. Soon another black yearling and a gray female yearling joined the first black. These would be three of the seven surviving siblings from the litter of eight born the previous year, the day the original Rose Creek alpha male was shot. The missing wolf had been killed by a delivery van over the winter. Of the surviving siblings, three were males and four were females, and one of the females was the lone gray of the litter. That was wolf 17. We did not spot any of the three pups sired by 8, nor did we see the two parents. The mother was probably inside the den with her pups, and 8 was likely out hunting. We started back down in midafternoon.

Ten days later, I saw one of the Rose Creek yearlings halfway between the pack's den and Slough Creek (pronounced *slew*). We thought perhaps 9 had moved her three pups to that area, but I did not see her or any pups. When I went back the next day, I spotted four of the yearlings, including the gray female, romping in a meadow on the other side of the creek. 17 did a play bow to one of her brothers, an invitation to chase her, and he obliged. She was faster than he was and ran circles around him. He nipped at her sides when she ran past him. When the gray got too far ahead, she ran back and circled around him, daring him to try to catch her. She seemed to take great joy in outrunning her brother and teasing him. Another of the female yearlings romped over to the black male and jumped up on his back. They playfully bit at each other. The gray ran in and all three played together.

After the play, one of the black yearlings saw a pair of sandhill cranes, probably parents near their nest, and chased them. The birds ran from the wolf at first, then flew a short distance. They landed, and when the wolf continued toward them, the pair flew across the creek. After that, the wolf pursued a small group of low-flying Canada geese. It seemed to be having a great time. Next the black chased a cow elk at top speed, caught up, and ran alongside her for a few seconds before veering off. The young wolf looked like a dog chasing a car just for the thrill of it.

The gray female chased a coyote, then went after twelve geese. When they flew off, she continued to chase them from below. After that, five cow elk came toward her. She dropped down into a stalking posture and charged at the last cow in line. When 17 caught up, the cow stopped to look at her. The wolf and the cow stood face to face, just a few feet apart. Then other cows came over and drove the yearling off. Later she stopped in a meadow, dug at a spot, and came up with a small rodent in her mouth, probably a vole. I saw her chewing it, then swallowing. A vole weighs only about two ounces. From a nutritional point of view, it would be like a person eating a few grapes or pieces of popcorn, something to take the edge off that hungry feeling, not a meal. Because these small rodents look a lot like mice, when wolves hunt them we say they are mousing. The wolves I have watched seemed to relish the challenge of hunting down these elusive little creatures.

I scanned the area and spotted four more black yearlings to the west. With the gray in the meadow and the two blacks who were now by the creek, that accounted for all seven of the Rose Creek yearlings.

That evening I saw 17 and a black yearling west of Slough Creek, where we thought their mother had moved her pups. A few minutes later a black pup, four to five weeks old, came into view and climbed up onto a boulder. That was my first sighting of a pup born to 8 and his new mate. One of the nearby yearlings looked at the pup and wagged its tail. A second pup, a gray, appeared and climbed up onto the same big rock. Another black pup came into sight. Then 8 came on the scene. The black yearling and one of the black pups ran to greet him. We saw what looked like the entrance to a new den, and 8 bedded down by it.

Later two black yearlings played with the three pups. One picked up an old bone and offered it to a pup, like it was giving it a toy. The other lay down beside a bedded pup, then got up to trail the third pup, who was walking around with a stick in its mouth. The yearling found another stick and romped around with it before presenting it to the pup. The pup let go of its stick and grabbed the new one, but soon dropped it. The yearling, trying to keep the game going, offered the stick to the pup again, all the while wagging its tail. The pup took the stick, bedded down, and chewed on it. The yearling then lay down in front of the pup and they sparred with their jaws. After that, the older sibling got up and found a small bone, which it brought back and dropped in front of the pup. The best moment came when the yearling wagged its tail and gently poked the pup with a front paw. As it left the pup, the yearling picked up the bone, tossed it into the air, and gracefully caught it in its mouth.

A black wolf with gray streaks on its coat then went to the pup. After rolling on its back under the adult, the pup wagged its tail and wriggled its paws in the air. Then it stood up and nursed on the wolf. That meant the adult had to be the mother, wolf 9. The other two pups ran over, and all three nursed as she stood over them. The pups had to stand as tall as they could on their hind legs to nurse. One of them placed a paw against its mother's hind leg to keep its balance. After five minutes the nursing session ended. We saw 9 nuzzle a pup. Then the three pups ran around. A yearling joined in their play, then brought them back to their mother. All of them played for some time, mother, yearling, and pups.

I had seen the alpha pair, three of the seven yearlings, and all three new pups sired by 8. It was a good day to be wolf watching, my best day so far that year with the wolves of Yellowstone.

EARLY THE NEXT morning, May 29, I reluctantly left the northern part of Yellowstone and moved into my temporary quarters in a dilapidated trailer at Old Faithful. Then I put on my ranger uniform and did a talk on wolves in an indoor amphitheater. That building was a challenge. It was built to show short movies about Old Faithful, short so that visitors had plenty of time to get out of the building to see the next eruption.

I started the season by presenting the slide show I used when speaking on wolves throughout the country and overseas. It normally lasted forty-five minutes. That was too long, for people got anxious about missing Old Faithful, the event they were really there to see. They would constantly check their watches, then get up and leave in the middle of the program. If an eruption took place during the early part of my talk, people that had watched it would drift in partway through the show. I had to figure out a new way of doing things. I gave up on the slide show and gave a five-minute summary of the wolf reintroduction program instead, then asked for questions. Park visitors always had plenty of questions about wolves. As people arrived after the geyser went off, it did not matter if someone asked a question I had already answered, because by then there were hardly any people left who had heard my earlier response.

One day I had a unique experience. I had my wolf pelt with me, draped over my knee, and I was seated on a stool while I spoke. A man came in with a big German shepherd, and they sat near the front. As I talked, I noticed the dog seemed scared of the pelt and the scent of the wolf fur. It cowered by its owner. After a while the dog settled down, and I forgot about it. Later I saw that everyone was suddenly paying much more attention to me. Several people were pointing in my direction and laughing. At that moment, I felt something wet on my right pant leg. I looked down and saw the dog urinating on the wolf pelt. He was showing that wolf what he thought of him, and I got included in the deal.

I was off on Mondays and Tuesdays and did not have to start work until late morning on Wednesdays and early afternoon on Fridays. I planned to drive up to Lamar early on Monday morning, stay over through Wednesday morning, then head back to Old Faithful. I would also drive up there on Friday mornings. I had to get up at 3:00 a.m. to make it to Lamar by first light at 5:00 a.m., but I was willing to do it to see the wolves. I moved north to Madison Junction when my trailer there was fixed up. That shortened my drive by sixteen miles each way, but I still needed to leave well before 4:00 a.m. It was forty-three miles to South Butte, where I was monitoring the Leopold pack, and another fifteen to the Rose Creek territory.

ON THE EVENING of June 6, as I looked for the Rose Creek wolves near Slough Creek, I saw a cow elk chasing two young grizzlies. The bears were probably siblings that had just left their mother. Then I realized that one of them had a freshly killed newborn elk calf in its mouth. The cow, who must have been its mother, broke off the chase. As the two bears were feeding on the calf, the Rose Creek alpha pair and the gray yearling came on the scene and charged at the grizzlies. Stopping a few feet away, the wolves snapped at the bears, but did not make contact. Two more members of the pack, both black yearlings, joined the first three. The wolves then worked as a team: two charged at the bears and drove them away from the calf while another ran to the carcass. The bears turned around and ran at the two wolves behind them. One swatted at a wolf with a front paw, but missed. Unable to deal with the harassment, the young bears left the area, and the five wolves took over the kill.

By June 10, the Rose Creek wolves were at Slough Creek most mornings, drawn by the concentration of elk, particularly cows with newborn calves. That day I saw 8 and five of the yearlings. One carried a stick in its mouth. The yearling tossed it in the air and caught it. When it threw the stick up a second time, it arced over the wolf's back. Leaning backwards to catch it, the yearling tripped and fell, then instantly jumped up and caught the stick before it hit the ground. It was a display of physical grace and agility worthy of basketball player Michael Jordan.

Later 8 came over, picked up that stick, and passed it on to another yearling. It threw it in the air and caught it just as gracefully as the other yearling had done. I called it the stick-tossing game and thought about how wolves had invented it millennia ago and how their domestic relatives, especially dogs like golden retrievers, love to play a similar game with their human companions.

A few days later, I watched as 8 and four yearlings teamed up to chase a herd of fifty cow elk that had a newborn calf in their group. Soon 8 was right behind the calf. The yearlings swung over to him, ignoring the surrounding cows and staying with the calf. It ran into a group of trees with the wolves right behind it. I got glimpses of wolves and the calf running back and forth. The four yearlings then ran out of the forest with several cows chasing them, followed a little later by 8 with the calf in his mouth.

A yearling joined 8 as he dropped the calf and lay down beside it. The two wolves fed for a while side by side until another yearling came over, and all three jointly fed on the carcass. Finally, the gray yearling arrived and 8 allowed her to eat as well. He could have kept the calf to himself or taken it up to the den and given it to his own pups, but he willingly shared with the yearlings he had adopted the previous fall. I later learned that not all wolves are as generous and sharing as he was. Like people, wolves have different personalities. Some are selfish and unnecessarily violent with family members and rival packs, and others are not.

In studies of human behavior, the traditional big question is: Nature or nurture? Is a baby born with a personality that will stay the same for his or her life or does the child's parental care and upbringing determine that personality? I did not know it at that time, but I would have a wolf case history to study in the coming years. I was destined to follow and document the entire life story of 21, one of 8's adopted sons. As he lived out his long life, I studied his behavior to see how it might mirror what he learned from watching his father. Would his personality and character be like 8's or would it be totally different?

The next day I found the seven yearlings unsupervised at Slough Creek. That gave me a chance to see how they operated without any parental oversight. They soon spotted a cow elk and chased her. She kicked back and hit one of them in the head with her hoof. When the cow reached the creek and ran into the water, the yearlings gave up. Later they chased a group of cows and calves. Like the first cow, the elk all ran into the creek, which was in flood stage. Three stopped at the edge of the water, but the gray female jumped in and swam after the two calves in the group. She soon reached the slower calf and grabbed it on the back of the neck. As she held on to it, she seemed unsure what to do next. The water was well over her head, and she did not have much leverage to handle the struggling calf. She let it go. The calf swam the rest of the way across the creek and joined its mother. It appeared uninjured. The yearlings were trying, but they still had a lot to learn about hunting.

10

The Battle of Slough Creek

B Y FAR THE biggest day of the summer of 1996 was June 18. I left Madison at 4:00 a.m. and got to Slough Creek at 5:20 a.m. I walked up Dave's Hill, a low rise just east of Slough Creek, and began looking for the Rose Creek wolves. A small group of other wolf watchers joined me. I spotted the Rose wolves on the west side of the creek: 8 and all seven yearlings.

The pack saw a herd of elk cows that had one calf in it. Immediately 8 charged at them. All eight wolves chased the herd. The calf was positioned in front of the cows, so they were between it and the wolves. I saw that 8 was gaining on the calf. Soon the calf tired and gradually fell well behind the lead elk. The cows changed direction, then ran back the way they had just come. The calf made those turns as well. The lead wolves, including 8, had to turn around and lost

THE BATTLE OF SLOUGH CREEK | 81

ground in the process. The cows and the calf ran right into the last few yearlings, the slower ones. Four of them converged on the calf and killed it. The rest of the pack ran in a few moments later.

After feeding, all eight Rose Creek wolves were bedded down by the calf carcass. Suddenly 8 jumped up and looked uphill to the west. I swung my spotting scope that way, expecting to see more elk with calves, but instead saw four Druid Peak wolves charging down directly at 8 and his family. That was my first view of the Druids, the wolves that had attacked 8's original family, the Crystal Creek pack. The big alpha male, 38, who had torn apart his metal cage, the one who likely had killed 8's father, was leading the charge. Behind 38 were three of the pack's adult females.

I looked back at the Rose Creek wolves and saw 8 charging uphill, straight toward the much bigger wolf. That put 8 between the enemy wolves and the yearlings. I thought of the stories I had heard about how his brothers had constantly picked on little 8 in the acclimation pen. Due to his size, he had probably never won a fight against any of his bigger siblings. But now, to protect the yearlings he had adopted and bonded with, he was willing to fight what looked like an invincible opponent.

To reach the formidable Druid alpha male, 8 had to run uphill. That meant he would be tired and possibly out of breath when he met his opponent, who was running down the slope. 38 would have all the advantages in the fight: he was much bigger and stronger, he was older and more experienced in battle, and he had already proven his combat skills by defeating and killing 8's father. Now he was running down the ridge straight at 8. What chance would 8, his father's smallest son, have against such a wolf?

Several thoughts flashed through my mind. Would 8 be exposed as an inadequate alpha male, one who could not fulfill the basic responsibility of protecting his family? Had 9 made a fatal mistake when she chose him to be her mate? As I watched 8 valiantly running uphill to confront the enemy, I thought of how some Native American warriors, facing impossible odds in battle, would charge forward yelling, "Today is a good day to die!" This was probably going to be the day of 8's death. The two alphas slammed into each other and moments later were rolling around on the ground as they fought. Both were gray, so I could not tell which wolf was winning. Then the fight was over.

Do you believe in miracles?

I saw one gray standing in the dominant position over the other male, who was on his back. The upper gray bit the other wolf at will. It took me a few more moments to realize that the victor was 8.

The Rose Creek yearlings ran up and joined 8 in attacking the Druid alpha male. After twenty seconds, 8 stepped back and let his rival go. The yearlings followed his example. We saw 38 jump up and run off, back uphill, with his tail tucked between his legs. The Rose Creek wolves chased him up the ridge. The younger wolves soon lost interest and stopped, but 8 continued after 38 at top speed all the way to the top of that ridge. They went over the crest and we lost sight of them. During that chase, 38 looked deathly afraid as he glanced back over his shoulder at 8, a wolf much smaller than he was, but one that had transformed into a fierce warrior when he needed to protect his family. That incident was like watching David beat up Goliath, then chase him up and over a mountain.

Later I thought about what the yearlings, especially the three males, must have felt when they saw their adopted father defeat the much bigger alpha male. I recalled a quote from an athlete who went into the same sport his father had excelled in: "A boy wants to be just like his dad. That's nature. That's how it works." I'm sure the male yearlings felt the same way after witnessing what 8 did that day: 8 was their hero, their role model of what a male wolf should be, and they aspired to be just like him.

Those of us on the hill, now that all the wolves were out of view, looked at each other with stunned expressions, then talked about the incredible sighting we had just witnessed. The victory of the undersized 8 against his huge opponent was the most amazing thing any of us had ever seen in the wild. A classic underdog, 8 went from being the smallest of all the male wolves brought down from Canada to being the champion of his region of Yellowstone.

Wolf 8 was the champion for now, but there would come a time when he would have to face a fellow champion, one who was bigger and stronger than the Druid alpha male and far more skilled in battle, one who never lost a fight. That day would come when 8 was an old wolf, way past his prime and suffering from many injuries and disabilities.

A few days later we got a mortality signal from one of the Rose Creek male yearlings. A crew went to the site, north of where the battle had been fought, and found his body. Signs indicated he had been killed by other wolves. We think he followed the scent trail of the Druid wolves, ran into them, and was attacked. That reduced the yearlings to six out of the original eight.

THAT SUMMER I spent a lot of time trying to figure out how 8 could have won that match. I eventually came to think that 8 might have remembered a time when one of his bigger brothers wrestled with him, pulled him down, and pinned him. On slamming into 38, perhaps 8 used the same move on his opponent, and that won the match for him. If that was the case, then the long-ago loss was something he had learned from, and he used that experience to win the fight.

There was a reason I felt it could have happened that way. In our neighborhood in Billerica, there were a lot of kids, from preschoolers to high school age. I decided I wanted to hang out with the toughest kids, a semi-gang of various ages. Soon after I joined them, there was a day when nothing was happening, and we were just sitting around on a lawn.

Phil, a high schooler, was the leader. Looking at the other guys, he pointed at Tommy, one of the younger kids. Anyone could see that Tommy was big and strong for his age. All of us stared at Phil, wondering what he was up to. This was years before the movie *Fight Club* came out. I say that for when Phil next spoke, he said, "Tommy, I want you to fight..." As he paused, all the younger boys in the group hoped that he was not going to point at any of us, but at someone older. Then Phil pointed at the smallest and youngest kid and said to Tommy: "Fight him." He was pointing at me. I looked over at Tommy and knew there was no way I could beat him, but I also realized that if I backed down, my time in the group would be over. I stood up, looked at Phil, then at Tommy, and said, "Okay."

While we called them fights, they were really wrestling matches. All of us watched the local pro wrestling shows on television. We had seen the holds the wrestlers used on each other and knew the rules for their matches. To win, you needed to pin your opponent for a count of three or put a painful hold on him and make him give up.

We came at each other and, as I expected, Tommy was much stronger than I was. But I somehow kept up with him as we tried to pin or put holds on each other. The match went on, and I was aware the other guys were surprised it wasn't over yet. After a time, Tommy put one of those holds around my neck that we all had seen on the television shows. I could not get out of it. He applied more pressure, but much less than he was capable of. I struggled to free myself, but nothing worked. Saying the code words to acknowledge a defeat, I told him, "I give," meaning I give up.

Both of us stepped away. We never spoke of that match again, but it had a big impact on us for we became best friends and stayed that way until I had to move away years later. As I think back on the incident, I suspect that Tommy respected my willingness to accept the match with him, despite our size difference. He could have taken me down and pinned me or forced me to give up in the first few moments of the match, but he gave me time to look good in front of the other guys and made it appear as though I was holding my own against him. Because I stood up and accepted that match, I learned something I otherwise might never have known. I had an unexpected talent for wrestling and naturally knew how to use balance and leverage on opponents. That gave me a lot of confidence and enabled me to stand up to bullies.

Years later, in college, I promoted dances for a time. I was the social chairman of our dorm and used our budget to hold dances. Figuring that a lot more guys would pay to come, I always held them in women's dorms, rather than in our all-male dorm. I hired the rock bands, told them what to play, and sold tickets at the entrance. Since it was a one-man operation, I also had to serve as the bouncer. One night, after coming back to the dance floor, I heard a woman calling out in distress and saw a guy assaulting her. I yelled at him to stop. He turned to face me, and that allowed her to run off. Enraged at being interrupted, the guy came at me and five of his friends followed him. He swung his fist at my head. I dodged just enough to make it a glancing blow. Without needing to think about it, I stepped forward and put the same hold on him that Tommy had used on me many years earlier. That shut him down. I kept him in the hold long enough for him to know he could not get out of it, then let him go. As the guy left the dance, I silently thanked Tommy for teaching me that move. That incident is why I think 8 could have had a similar experience with one of his big brothers and used a move his brother had taught him to beat his much bigger opponent.

11

The Games
Pups Play

I DID NOT SEE the Druid wolves again until July 1. After their big alpha male lost the fight with 8, they stayed in Lamar Valley, well away from the Rose Creek wolves. Doug told me the white alpha female had been driven out of the pack by 40, the largest and most aggressive of her three yearling daughters. By August, 39 was a hundred miles north of the park.

After her mother's departure, that domineering daughter took over as the new alpha female. The pack now numbered four: the alpha male, the new gray alpha female, and her two black sisters. 40 ruled the pack ruthlessly, like Queen Cersei from *Game of Thrones*. Her two sisters, 41 and 42, were deferential and never stood up to her. The two blacks got along well together. I once saw them walking along side by side as they jointly carried a big elk antler. When they interacted, 42 usually seemed more dominant. Occasionally, 41 pinned her sister, but those times may have been just play. I saw only one squabble between them, and 42 pinned her sister twice during it.

That July day, I saw 38 and 40 do six double-scent marks together, which confirmed they were the alpha pair. On each of those occasions the male lifted a hind leg up high and urinated on a bush or tree, then the female came over, sniffed the site, partly lifted a hind leg, and marked over his raised-leg urination with her own flexed-leg urination. The two subordinate females did not do any scent marking.

We watched the Druids from 7:16 a.m. through 9:41 p.m., nearly fourteen and a half hours. They rested from 10:12 a.m. to 8:42 p.m., then got up and killed an elk calf. The rest period comprised 73 percent of the time we had them in sight. Most of the remaining time was spent traveling and looking for prey. Those figures do not capture the complete picture because wolves are also active after dark. They have excellent night vision and can also detect hunting targets by scent. In later years, using data collected from special GPS-equipped collars, Wolf Project staff found that wolves are most active during the twilight hours just before and after dusk and dawn.

THE DRUIDS GOT harder to find after July 1 as the elk migrated to higher elevations, so I returned to South Butte and resumed my observations of the Leopold wolves. On July 8, I saw the Leopold pack's three pups for the first time. Two were gray and one was black. The adults had moved them to a rendezvous site about a mile southwest of the den forest, and I had a good view of that area from South Butte. A rendezvous site serves as a convenient aboveground place for the pups to hang out when the rest of the pack goes off on hunts. There are often old coyote or badger burrows there for the pups to explore and hide in if bears or other predators come around.

On that day, and every other day I watched them, the pups played together for lengthy periods. They chased and wrestled with each other constantly. I saw the black pup pick up a small stick, run over to one of the two gray pups, and show it the stick as though it was daring the gray to try to get it. The gray chased the black, then the black turned around and chased the gray. That was the same game of catch me if you can I had seen the Crystal yearlings play the previous summer.

I noticed much of the play involved sparring with their jaws. Two pups would get face to face and bounce back and forth, looking for an opening to nip at each other. Like boxers, the pups did a lot of feinting, bobbing, and weaving. They made jabs at each other with open mouths. The game of sparring trained the pups for serious fights with rival wolves when they became adults.

Another favorite pup game was tug of war, which could involve a stick, a bone, or a section of hide. I once saw the alpha female play that game with one of the gray pups over a chunk of meat. 7 also liked to play catch me if you can with the pups and took turns either chasing a pup or pretending to run from one. When the pursuing pup caught up with her, mother and pup would wrestle and spar with their jaws. The alpha female enjoyed playing so much she often leaped up and down in front of a pup and did play bows to keep the game going. She also played by herself. Once she got a small piece of meat, ran back and forth, leaped up, and tossed it in the air. She did that six times and caught the meat almost every time. Then she ran around in circles just for the fun of it. On another occasion I saw her trying to catch her own tail.

The alpha male also regularly played with the pups. I saw a gray pup go to him while he was lying down, sniff him, and put a paw on his shoulder. The big male jumped up and ran off. The pup chased him but tripped and tumbled. The father wolf waited for the pup to get up and run over to him. The two wrestled, then he ran off and dodged the pup when it came after him. Later 2 bowed down and let the little guy hit him with a front paw and nip at his face. At times the male was tired and needed to rest for the next hunt. During one of those moments, the three pups ran over to him, expecting a play session. Two of them climbed up on his back. That got him up, but rather than play with the pups, he quickly walked off. The pups followed, but soon got distracted. He ended up hiding from them behind a clump of tall grass and got his nap.

The black pup was male and seemed to be the ringleader and instigator of many of the games. The two grays were female. That first day I saw the black pup play the game I called ambush. He ran at the two grays, got past them, dropped down into the grass, then jumped on them when they ran to his hiding spot. The black pup would often see one of the gray pups bedded down, then run over and pounce on her. When he was alone, the black could invent his own games to entertain himself. A common plant in that area was mullein, which grows in the form of a tall stalk. He would run over to one, jump up, grab the top of the stalk in his mouth, and pull it down. When he let it go, it would spring back to vertical. The pup would do that trick over and over.

I noticed that the mother usually kept track of where the pups were. If one wandered too far away, she would run over and intercept it, then divert its attention by getting it to play with her. With grizzlies, black bears, and coyotes in the area, she needed to monitor her pups and not allow them to get too far away. She was a young mother, and these were her first pups, but she instinctively knew how to care for them.

By the time the pups were three and a half months old, I saw them following scent trails at the rendezvous site. I watched the black pup trail the scent of his parents after they left on a hunt, like a bloodhound following the trail of an escaped prisoner. The pups were learning how to be wolves.

By mid-August, the four-month-old pups were traveling with their parents well away from the rendezvous site. I got up to South Butte early on August 19 and saw the family of five moving my way, with the alpha male in the lead. The pups played together as they followed the adults. At times, a pup would go out in front of the adults and lead. When the pack approached a huge bedded bison, the pups hesitated and ended up circling around it as their parents waited for them.

The mother was in a playful mood that day. She romped with her mate and jumped up at him. At times 7 ran circles around him. They would get face to face, rear up, then spar with their jaws. Later he ran ahead, dropped down into the ambush position, then jumped up and charged at her when she approached. I wondered if the parents were in such a good mood because they anticipated the pups would soon be able to travel with them full time throughout their entire territory.

MY SUMMER JOB in the Old Faithful and Madison areas ended in early September, just after Labor Day. I moved out of my government trailer and rented a room in Gardiner, the small town north of Mammoth, and went out to Slough Creek and Lamar Valley every day until the middle of November.

When I found the Druid wolves on my first visit back in Lamar, I saw the new gray male that had joined the pack in August. Wolf 31 had been brought down from British Columbia as a pup with other wolves in what would be called the Chief Joseph pack. They had been put in the Crystal Creek pen, then placed in a temporary pen and released on the west side of the park, where they established a territory. 31 later left his pack, traveled east to Lamar Valley, and somehow managed to join the Druids. Although alpha males normally drive off potential rivals, 38 seemed to have no problem accepting this new wolf into his pack.

We later discovered through DNA testing that the two males came from the same pack. In Canada, the big Druid alpha had been captured as a lone male and his pack of origin was unknown. He had probably left his original family after 31 was born, because the two wolves acted like they knew each other. When I saw them play together that day, they seemed very friendly. A few weeks later, I watched as the newcomer played with all three Druid sisters. At one point he ran circles around the two subordinate sisters. 38 came over and playfully chased him. All that indicated how well 31 had integrated into the pack.

The acceptance of the new male into the Druid pack made me think about wolves' ability to find relatives and acquaintances. When the young male had been in the Crystal Creek pen the previous winter, he had probably heard the Druid wolves howling in the Rose Creek pen five miles away. We think wolves can identify other wolves they know from the sound of their howls, like people recognizing the voices of friends. That means he had probably recognized the howling from his relative. After his group was released on the west side of the park, 31 came back by himself to Lamar Valley to reunite with 38.

That fall I noticed a different side of 38. I saw the other Druid wolves playfully chase him. After running off, he came back, dropped into the ambush position, then jumped up and pounced on one of the wolves when they all ran to his position. After that, he ran off, inviting them to chase him. When they failed to pursue him, he ran back, spun around when he reached the others, and ran away once more, trying to keep the game going. I had classified him as being violent and aggressive, but now I saw that there was also a playful side to his personality.

I HAD NOT seen any Rose Creek wolves since June 19. On September 17, I spotted ten pack members north of Slough Creek. I had the same count on October 22. Of the original eight yearlings, just five were left and four of them were females. One of the males had been hit by the delivery truck, one had been killed by the Druids, and a third one had left the pack. The sole remaining male was 21, the last pup Doug had pulled out of the den that fateful day the previous spring and the one who had guarded the family when his mother and his siblings were returned to the acclimation pen after the alpha male was shot. If anything happened to 8, he would be the pack's next alpha male.

Starting on November 2, I failed to get 21's signal when I checked on the pack. That is a common time of year for young males to disperse. The February mating season was only a couple of months away, and I thought he might be searching for an unrelated female to pair off with. But he returned from his walkabout nine days later, and the pack once again numbered ten.

It snowed around that time and the Rose wolves played in it. Three of the yearlings repeatedly slid down a steep slope covered with snow. This was a new game: snow sliding. Winter was coming to Yellowstone and thousands of elk were migrating down into Lamar Valley, where the snow was not as deep as it was in higher country. I counted 545 elk in one herd.

All the time I put into watching the Yellowstone wolves in 1996 climaxed with a sighting of the Leopold pack from South Butte on November 11. It was an event that perfectly demonstrated how in a wolf family individual members cooperate to accomplish a dangerous mission.

That evening I saw the alpha pair chasing a big cow elk. Two of their pups ran along behind them, the black and one of the grays. The alpha male was leading. He caught up to the cow and grabbed her hind right leg. She kicked back at him with her other hind leg and hit him several times with her hoof. Despite those powerful kicks to his head, 2 held on. The cow was now slowing down as the big male acted as a drag. The alpha female caught up with them. 7 got out in front of the cow, turned around, then leaped up and grabbed her throat, the classic finishing move of a wolf.

The alpha pair were working together like two defensive football players tackling a much bigger opponent. The alpha male's bite on the hind leg was not doing much damage to the cow. His role was to hold on to her so 7 could get out in front and deliver the killing bite to the throat. As the two adults struggled with the cow, the gray pup ran in, but did not know what to do. Then the black pup arrived and without hesitation bit into the side of the elk. The alphas and the black pup wrestled the cow to the ground and soon she was dead.

The black pup got credit for the assist that day. He was seven months old at that time, about equivalent to a seven-year-old boy. When I had first seen him in early July, he had been just three months old and knew nothing of hunting. Now, just a few months later, he was a huge asset to his family. He was like a third-string rookie on a football team, jumping into the middle of a game and helping the team win, despite his inexperience.

Two days later, I went out into the park as usual in the morning, then came back, packed my belongings in my van, and drove to Big Bend for my fourth winter.

12

Den Troubles
for the Rose Creek
Wolves

IN THE FALL of 1996, the supervisors in Mammoth agreed that placing me down at Madison and Old Faithful, well away from where the Rose Creek and Druid wolves were being seen in the northern section of the park, had not been beneficial to park visitors, and they told me I would be living at Tower again for the summer of 1997. On May 13, I unloaded my belongings in a government trailer there, then went out in the evening to look for wolves.

I had heard that the Rose Creek alpha female was denning on the south side of Little America, an area about two miles east of Tower. She had a litter of three black pups and four grays. The name of the general area originated in the 1930s when a Civilian Conservation Corps camp was located there. During the bitterly cold winters, men stationed in the camp claimed it was as cold as Little America, a research station in Antarctica.

I spotted 9 and a gray pup at the reported site, five hundred yards south of the road. Both went into trees farther to the south. Then 8 came out of the forest, and three black pups ran to him. They mobbed him, licked his muzzle, and he regurgitated meat to them. The pups lowered their heads and fed voraciously. A raven landed and walked toward the pups, hoping to steal some of the meat, but 8 chased it off. The area where I spotted the wolves was the pack's rendezvous site. The pups had been born in a burrow in the trees a short distance to the east.

One of 9's adult daughters sired by 10, the original Rose Creek alpha male, had been bred by 8 in February. The young mother, wolf 18, now had a litter of pups at the pack's 1996 den site near Mom's Ridge, on the other side of the park road and three miles north of the alpha female's current den. The road was not the only obstacle between the two sites. The wolves also had the Lamar River to contend with. The Wolf Project staff thought 9 had gone out on a hunt toward the end of her pregnancy and was south of the river and the park road when her pups were about to be born. Having no time to get back, she had her seven pups there. The two dens were close enough that the Rose wolves at one site could hear pack members howling from the other den and call back to them.

8 and the remaining male yearling, 21, stayed at that new site and helped 9 with her pups. 21 was now just over two years old, about twenty or twenty-one in human years. He had been with 8 for eighteen months, about three-quarters

of his life, and the two males had a strong bond. The three pups born in 1996, now yearlings, were based at Mom's Ridge with the new young mother and her pups. The river between the two dens was in flood stage due to melting snow, and it was dangerous for the wolves to swim across.

There had been a third den site, located just east of the Slough Creek Campground road, three miles from the other dens. Another female, wolf 19, from the original Rose Creek litter, also pregnant by 8, had given birth to four pups there. Soon after their birth, her collar sent out a mortality signal. A Wolf Project crew member found her body on April 19 and determined she had been killed by other wolves. The Druid wolves had been seen near her den and therefore were the prime suspects. Her four pups were found dead of starvation three days later. As the Druids had killed her brother after the Battle of Slough Creek, that meant that they were now responsible, directly or indirectly, for the deaths of six Rose Creek wolves.

Both 8 and 21 had originally gone to all three dens and brought food to each mother. That had been a very inefficient situation for the pack. After the death of 19 and her pups, the two males still had to go back and forth from 9's den in Little America to her daughter's den near Mom's Ridge, despite traffic on the road and the hazardous river conditions.

That day, my first day back in the park, both 8 and 21 were seen regurgitating meat for the pups at Mom's Ridge. The next morning, I saw the two males at the alpha female's den, where all seven pups were frantically begging them for a feeding. 21 had become a big male. He seemed larger than 8 and looked like he could be his adopted father's bodyguard even though he still acted subordinate to him. His loyalty reminded me of how rescue dogs remain devoted to the person who took them in. They never forget the kindness shown to them.

Throughout his life, that was the word that defined 21: loyalty. I recalled how his biological father had stayed by the Rose Creek pen after walking out through the open gate, patiently waiting for the two females to join him, despite his likely assumption that people would return to the area and recapture him. Like his son 21, loyalty to his family was a defining element in 10's character.

As I watched, 9 came over and joined the pups and two adult males. Her black coat now had a lot of gray streaks, due to aging and perhaps the stress of being a mother. She bedded down and a pup climbed on top of her. Soon she and 8 went into the trees, and all the pups followed. I saw a large boulder at the edge of the trees up in that area. There were burrows under it the pups could explore and hide in if a bear came along. Later that day I saw both a grizzly and a black bear near the site. When the black bear approached, 9 ran out and chased it off. At the end of the evening, I saw pups following 21 around like Cub Scouts marching behind their troop leader.

Bears and coyotes were a constant threat to the pups that the adults had to deal with. Linda Thurston and her den study crew were monitoring the den when they saw a black bear get to within twenty-five feet of the pups before 8 ran in and chased it up a tree. Another time, 21 drove off a mother bear with three cubs. On another occasion, two coyotes approached the site and 9 chased them away.

One day I spotted 9 coming out of the trees with the pups. She lay down on her side and nursed them. Later, 21 returned to the den and the pups mobbed him, trying to get him to regurgitate meat for them. I noticed that there were only six pups pestering him. The seventh pup, a gray whose coat had a tannish tone, stayed uphill by the big boulder. I wondered why that pup was not down with the others. 21 lay down and played gently with one of the pups. The five others ran over to get a feeding and play with 21, at which point he got up and walked off. The six pups switched to playing with each other and let him go. I later saw 21 watching the pups from the edge of the trees. He looked like he was on security duty, scanning for intruders.

Early on May 17, I found both 8 and 21 in Lamar Valley, feeding on two different fresh carcasses. The two males must have gone out hunting during the night and killed both elk. I watched 21 carry off a large piece of meat in the direction of the den. 8 followed. When I lost sight of both males, I drove back to the den area and saw 21 coming in with the meat. He went into the trees, where the pups were probably resting. The alpha female ran out from another section of the forest wagging her tail in excitement, then disappeared where I had lost 21. Within a minute, she reappeared by the big boulder with her son's meat and put it on the ground.

Six pups ran out of the forest to greet 21, who had also come out from the trees. The seventh pup, the one with the tan shading, hung back as he had done before. I thought he might be sick or injured, and later I saw he had trouble walking and fell frequently. Because his movements were limited, the pup stayed alone uphill by the boulder, watching the other pups and 21, like a sick child looking at other kids playing. A few minutes later, 21 walked off from the six pups and trotted up to the seventh one. He sat beside the pup for some time, a big brother with his little brother, then returned to the others. That was a profound moment to witness. 21 would have been tired from the nighttime hunt and from carrying the big piece of meat to the den. He needed to rest for the next hunt. Despite that, he had noticed the last pup alone uphill and had gone up there to hang out with him.

I have told thousands of people that story in my talks over the years. I then ask if they ever had a time when they were a kid when they came home sad or depressed, then had their dog run over, seem to understand their mood, and want to hang out with them. Nearly everyone said yes to that question and nodded when I asked if that had cheered them up. Dogs know what it is like to feel sad, lonely, neglected, or sick, can sense those feelings in humans through our body language and facial expressions, and can help us feel better just by choosing to be with us. They want to be our friend when we most need one.

If people appreciate that aspect of a dog's personality, they should understand that it comes from dogs' wolf ancestors, from a wolf like 21, who noticed that pup and went to him, out of empathy for his situation. If a sad or sick child can be cheered up by having a dog come over and want to be with them, then surely 21 cheered up that pup when he spent that time with him.

Seeing 21 go to that sick pup reminded me of another story. Years ago, before I started working in Yellowstone, I did some joint programs with Kent Weber and Tracy Ane

Brooks of Mission: Wolf, a Colorado nonprofit organization that cares for unwanted or abandoned captive-born wolves. I would present a slide show on wolves, then introduce Kent, who spoke about his wolf sanctuary. The climax of each event was the appearance of one of Kent's wolves who had been socialized to be around crowds of people. As the audience watched, Kent walked the leashed wolf around the room.

After one of our joint appearances, Kent told me a touching story. He often did talks in elementary schools, sometimes to as many as five hundred kids. For school programs, he usually brought one particular black female wolf named Rami. Kent would ask the children to stay seated and not reach out to her, but added that it would be all right to pet her if she voluntarily came to one of them. As Kent walked the wolf around school auditoriums, Rami often picked a child, went to him or her, and allowed that kid to stroke her.

Trying to understand what was going on, Kent asked teachers if there was anything special about the child the wolf had singled out. Usually the teachers told him the boy or girl was the one most picked on and bullied by the other kids. Rami sensed the distress those kids were going through and reacted by going to them and having a moment of friendly interaction. Each time that happened, it not only cheered up the chosen child, but since it took place in front of the entire school, it permanently upped the status of the boy or girl.

There is another story that illustrates a wolf's capacity for empathy. At Wolf Haven, a sanctuary for captive-born wolves in Washington State, a couple with a crying baby walked up to a wolf enclosure. A female wolf who was bedded down heard the crying. She got up, dug out a food cache, went over to the fence, and pushed a piece of meat toward the baby.

I thought about where 21 got his sense of empathy and realized it probably came from 8. When 21 and his siblings spent the first six months of their lives in the Rose Creek pen, they had no father or any adult male wolf in their life. Then 8 arrived, made friends with the pups, and volunteered to serve as their father. From that moment on, 8 was 21's role model for masculine behavior. Later 21 saw that tan pup isolated from the other pups, and as 8 had done for him and his siblings, he did something to help.

Over the years I have been asked by the Make-A-Wish organization to take sick children out to see wolves. Those days are always the best ones of the year for me, when I get to help a boy or girl have a good time and forget their troubles. I do that in remembrance of what I saw 21 do that day.

That evening I saw 21 bed down by the big boulder. Suddenly, he jumped up and ran off to drive a bear away from the den area. Later when 8 and 21 were out with the six pups, I noticed the seventh pup slowly walking around uphill. It was promising to see him moving. Then 21 left the rendezvous site to chase a herd of about a hundred elk. He picked out a cow and killed her by himself.

When he came back to the den, the six pups ran to him. The lone pup wagged his tail, then in a playful prancing gait, started to move downhill. He even climbed over a log. Soon he was with 21 and the other pups. All of us watching were overjoyed. The seven pups encircled 21, then one of them ran off with a piece of meat, proof that 21 had gulped down meat at his kill, rushed back to the pups, and regurgitated it to them. Later that morning, 21 went back to the carcass and used his prodigious strength to carry a heavy section of it back to give to the pups.

The pups loved playing with 21. He was like a favorite uncle in a human family. When he bedded down, they would rush over and paw at his face, lick him around the mouth, and climb up on his back. If he got up and walked off, they would run after him. He tolerated their attention in a good-natured manner.

One morning I saw 21 jump up and run over to greet his mother, all the while wagging his tail. The six pups ran to them, and the lone pup shuffled downhill toward them. 21 saw that pup and ran to greet him, his tail still wagging. The pup made it down to 9, and all the pups nursed for five minutes. The tan pup was definitely getting better, and I thought it was partly due to the encouragement he was getting from 21.

I was becoming more and more impressed with 21. He was the only young adult that had stuck with his parents at the new den and helped with the pups. He kept bears away, went hunting by himself, repeatedly made kills and brought back meat for the pups, played with the healthy pups, and made a special effort to tend to the sick one. Research on wolf predation in the park by Wolf Project staff, including Dan MacNulty, who went on to be a professor at Utah State University, found that two-year-old pack members tend to be the best hunters because they are at the peak of their physical abilities. 21 was exactly that age. He had also begun to do raised-leg urinations at sites marked by the alpha pair, another sign of his maturity.

Just a year older than 21, three-year-old 8 was also a proficient hunter. One morning Linda's crew saw him kill an elk close to the pack's northern den. He then swam the river, crossed the road, and made a second kill by the den in Little America. Kevin Honness, who was doing observations for Linda's den study, recorded a hunt where 8 grabbed the throat of a cow elk while 21 bit into a hind leg. The elk was so tall that she lifted 8 off the ground, but he held on. She soon collapsed on to the ground. 8 shook his head side to side as he continued to hold on to her throat while 21 bit her back. When she tried to get up, 21 switched to her throat, got a holding bite on it, and helped 8 hold her down. She stopped moving five minutes after the wolves made first contact with her. 21 later tore out a choice part, carried it directly to his mother, and gave it to her.

One morning I saw 8 returning to the Little America den area after a hunt. He was mobbed by all seven pups, including the tan one. Six pups gulped down the meat he regurgitated, while the seventh leaped up to his face, trying to get another feeding. Then 9 arrived with more meat. Some of the pups fed on the meat while others nursed. After that, I saw her licking the little tan pup, singling him out for more attention than the others.

ON THE EVENING of May 24, I saw about twenty cars pulled over on the side of the road north of the Little America den area. I had not spotted 9 or her seven pups for three days. A visitor told me he had seen her cross the road from north to south with one pup, then move toward the den farther to the south. After going a few hundred yards, she had stopped and looked back. The visitor turned around and spotted more pups north of the road. Those pups saw cars and people between them and their mother and ran off farther to the north. By then 9 was howling to get them to come to her. I figured she had been attempting to move her pups to the original den at Mom's Ridge, had failed to get them across the Lamar River, and now needed to get them back across the road and to her den in Little America.

I talked to more people on the scene. Everyone was out of their cars, looking for the wolves, especially the pups. The visitors did not realize they were positioned directly between the mother and the pups still north of the road. I was on duty and in my ranger uniform and had to quickly figure out what I could do to help. The critical issue was getting the mother and pups back together.

I went to the people in each car, explained the situation, and asked if they would be willing to help the mother wolf get back to her pups by driving down the road, thus moving away from her direct route to the pups. Everyone I talked to understood the situation and volunteered to move off. I also drove away. I was very impressed with the visitors' response. They had all come to the park in hopes of seeing wild wolves but were willing to give up that chance when they heard their presence was prolonging the separation of a mother from her pups. They regarded the welfare of the wolf family as more important than their desire to see the wolves.

I later got reports that 9 had been seen with several pups north of the road heading for the Lamar River. That seemed to confirm my guess that she wanted to get the pups to the Mom's Ridge den where the rest of the Rose Creek pack was based. But the river was an even greater obstacle than the road. The water was high due to a record snowfall that winter, and the current was treacherously fast. It would be hard for her to find a safe place to cross with the pups.

Two days later I saw 9 and five other adult Rose wolves, including 8 and 21, north of the river and several miles west of the Mom's Ridge den. She must have crossed the river on her own and left the pups behind. The six wolves were chasing a group of elk. One cow stopped, looked back at the wolves, and stood her ground. When the lead wolves reached her, she panicked and ran off with the wolves in pursuit. She must have had something wrong with her, for they easily caught up and ran alongside her before leaping up and biting at her. Then they all ran behind a knoll, and it looked like they killed her there.

Later that day I spoke with Jason Wilson, a Wolf Project volunteer. He had been watching the northern den that morning from Mom's Ridge and had seen 9 arrive. She was the one who had organized the hunting party, and all the wolves except the new young mother, wolf 18, had gone with her to the west, where I saw them chasing the cow. Jason saw the wolves return to the Mom's Ridge den later in the day. The two mothers picked up 18's pups and carried them one by one to a new den site a mile away. In later denning seasons, we saw mothers make similar moves, but we usually did not know why they did so. Perhaps the den was caving in, a grizzly was frequenting the area, or the local prey animals had migrated away from the original den site.

The next morning, I saw 9 back near her den in Little America. Four days later, I got a report that she had been seen with a black pup and a tan pup, trying once again to cross the road to the north. She got across, but the pups turned back when people stopped their cars. The mother then recrossed the road and took the pups back into the trees to the south. By mid-June, three of her pups had been spotted north of the road, but that still presumably left four south of the road. She was seen repeatedly trying to encourage the pups that had crossed the road to swim across the river, but each time they balked. I thought it was a good decision on their part, for the fast current would surely have swept them away. On June 17, Linda's den crew got signals from 8 and 21 in that direction, indicating they were on the north side of the road with the pups who had refused to risk their lives in the fast-flowing waters of the river.

By then 9 seemed to have abandoned her den in Little America and was based at the den at Mom's Ridge. In her thesis, Linda wrote that 9 left that den every day and swam across the river to the pups trapped between the road and the river. Sometimes other adults went with her. After each visit, she headed back across the river to the northern den. The pups followed, but always turned around when 9 went in the water. An elk carcass was later seen near the pups' location and presumably they fed on it.

After a few weeks there were no more sightings of the pups by the river, and 9 stopped traveling back and forth. Later the remains of one pup were found near the water, but the cause of death could not be determined. The high count of pups at the main den was eleven, but two were lost, and so by late June the number dropped to nine. We did not know if any of the surviving pups were born to 9. That also meant that the fate of the tan pup, the one that 21 had tried to help, would never be known.

That spring there was a change of staff at the Wolf Project. Mike Phillips left to serve as the director of the Turner Endangered Species Fund, founded by Ted Turner, and Doug Smith took over as project leader.

13

Druid and
Leopold Pups

THAT SAME YEAR, 1997, the Druids also produced multiple litters, with two of the three young females raising pups. Unlike the pups in the Rose Creek pack, all the Druid pups were at the same den site, located in the forested hills north of two parking lots known locally as Footbridge and Hitching Post. Druid Peak was just uphill from that site.

On May 22, I got my first sighting that year of wolf 39, the original alpha female who had been driven out of the pack by her gray daughter, 40. She had returned earlier in the month from her long solo trip to the north. Linda Thurston documented that she was helping to care for her grandchildren. I saw 41, the lower ranking of the two black sisters, south of the den area and noticed that she had distended nipples, indicating that she was nursing pups. As far as I could tell, the pack's four females were getting along well, or at least they seemed to be as I watched them feed next to each other on a moose carcass.

The new alpha female apparently had no problem with her mother returning after being away for nearly a year, despite the fact that she had been the one who had driven her out of the pack. Perhaps 40 realized the family needed all the help it could get feeding and raising the pups. 39 appeared to understand that, despite being the oldest female in the pack and the mother of the three sisters, she was now at the bottom of the female hierarchy.

In late June, I finally spotted the Druid pups in a clearing known as the Diagonal Meadow. There were three blacks and two grays. Genetic tests later determined that the pups had been born to 41 and 42, rather than to the alpha female. 40 had been seen breeding with alpha male 38, but no pups resulted from their mating. I wondered how she was dealing with the fact that she did not have pups, while both of her lower-ranking sisters did.

I often walked from the Footbridge lot and hiked up a slope called Dead Puppy Hill. It got that name in the summer of 1995 when the Crystal Creek wolves dug out a coyote den there and killed several pups. From that hill we had a better view of the den area, especially a marsh between the den forest and the road. The pups discovered that marsh in early July and went to it repeatedly. It was there that they learned how to hunt voles and began to feed themselves.

The pups became so obsessed with vole hunting that they sometimes ignored incoming adults who would have regurgitated elk meat to them. The pups developed a technique where they would listen for a vole rustling through the meadow grass, then leap up and pounce on the spot with their front paws. Not always, but often enough, a pup would pin a rodent and then gulp it down. When they were full or bored, the pups switched from catching voles to playing with each other. The adult wolves often rested on the hill above the marsh where they could look down and check on the pups. Grandmother 39 spent the most time supervising the pups, and she often followed them around as they hunted voles, as though she was evaluating their hunting techniques.

On many days that summer, I watched the Druid pups in the morning, then went to South Butte later in the day to study the Leopold pups. The Leopold alphas, 8's brother wolf 2 and the former Rose Creek female wolf 7, had five new pups, and the family was as playful and interactive as ever. Keeping tabs on two packs tired me out, but I could not pass up the unique opportunity.

On the morning of July 10, the Druid pups got into a difficult situation, and I watched as the adults worked together to keep them safe. I saw the five pups and four adults in the Diagonal Meadow. 39 then moved downhill toward the road, and the alpha pair and the pups followed her. The three adults crossed the road to the south. When the pups reached the road, they walked back and forth on it, sniffing the interesting scents on the pavement. I spotted the alphas well to the south by Soda Butte Creek. They looked back, saw the pups milling around on the road, and ran back. Fortunately, no cars were coming.

The pups were still on the road when 38 reached them. Cleverly, he trotted past without greeting them and went back up the hill toward the Diagonal Meadow. All the pups followed. 40 stayed on the road until all the pups were off it, then trailed them up the hill. The grandmother ran back and headed up the hill after the pups. 41, mother of some of the pups, had not gone downhill with the other adults, and she was standing at the top of that slope. When the other adults and pups reached her, she led the pups farther uphill, away from the road. I was impressed with how the adults teamed up to resolve the crisis.

For the next few weeks, the adults led the pups farther and farther away from the den along routes that did not involve any road crossings. In early August, they brought the five pups three miles to the west, all the way to Rose Creek, where the acclimation pen still stood. After that, the family repeatedly traveled back and forth from the den forest to that drainage. They also explored the high ridges above that area.

In mid-August, 39 led the pack down to the road from the den. The alpha female and one of her sisters followed, as did the pups. The adults crossed the road, and this time the pups trotted after them. We saw the three females wade the shallow creek and continue south. One pup also crossed the creek, but the other four turned back and ended up stuck between the road and the creek. The pup that went through the creek was soon up on Dead Puppy Hill, where it started to howl. The four pups to the north howled back, ran to the creek, balked at going into the water, and fled back to the north.

The big alpha male saw the problem and returned to the north. First he regurgitated meat for the four wary pups, then, after they had fed, he led them down to the creek and crossed to the other side. One pup waded the creek and joined him on the south bank. That left three pups whining on the northern side. When the father wolf continued south, those last pups overcame their fear, crossed the water, and joined him. Already over their trauma, the pups played with each other.

That was the second time the Druid pups had got in trouble trying to cross the road or creek. Both times 38 had led them to safety. He seemed more capable of dealing with those problems than the other pack members, even the two mothers. He projected a calm confidence that helped the pups know things would be all right if they followed him. I was seeing an impressive new side of this father, a wolf I had disliked for his involvement in attacks on the Crystal and Rose wolves.

Now that they were all safely across the creek, the pups led to the south, excited to explore new country. They soon joined up with the fifth pup who had been howling for his siblings. The pups continued to lead with the alpha male trailing them. Soon most of the adult females came in from the south to join the group. They all ended up on the western side of Mount Norris, south of Hitching Post. The pups explored the area and seemed comfortable staying there. We called the area the Norris rendezvous site.

Two days later I saw some of the Druid adults at the Rose Creek den in Little America, thirteen miles to the west. The wolves soon left that area and traveled to the top of Specimen Ridge, a high ridge up to 3,000 feet above the south side of Lamar Valley. The next day, I got 21's signal at the Little America den site. His signals soon indicated that he was following the scent trail of the Druids up and over Specimen

Ridge. It normally would be dangerous for a single wolf to follow the trail of an enemy pack, one that had previously killed two of his siblings, but 21 had grown into a big and powerful adult and looked like he could take care of himself. Perhaps his intent was to check out the Druid females and draw one off as a mate. 42 was the only single female in the group he was trailing.

In late August, the Druid adults took the pups west and eventually settled into what we later called the Chalcedony Creek rendezvous site. That involved getting the pups across the Lamar River. On the trip, most of the adults got well ahead of the tiring pups, but 39 stayed behind with them as the pups followed the pack's scent trail. The pups seemed to especially like this site, for it had a lot of old coyote den tunnels they could explore.

OF ALL THE packs I had been watching, the Leopold wolves seemed to have had the easiest time that summer. I thought back a couple of months to June 23, when I had walked up to South Butte for the first time that year to look for them. I had spotted mother wolf 7 bedded with her five new pups. The two gray yearlings were with them. I later saw alpha male 2 and the black yearling who had impressed me the previous fall with his involvement in the elk hunt when he had been a pup. After resting, the pack got up and traveled. I saw the alpha male playfully romp around by himself, then toss a bit of fur in the air, just as he and his brothers had done in Lamar back in 1995. He had a big family and a lot of responsibilities, but he still liked to play, even if it was by himself.

I watched as the Leopold alpha pair went out of sight in a ravine. A herd of seventy-five elk cows and calves appeared, and 7 ran out and charged at them. She targeted a slow-moving calf, but several cows cut her off. Then her mate ran in and the pair closed in on another calf lagging behind the herd. The wolves made the kill. After feeding, they went after another sluggish calf, but three cows drove them off. The black yearling ran in and joined the alpha male on a chase of a different calf. As that calf fell farther behind, 2 ran in and grabbed it.

The Leopold alphas worked hard to feed their pups. Often after making a kill, they had to confront black bears who were intent on stealing it from them. One day I watched as the pair chased a black bear off one of their kills and up a tree. The bear looked down at the male, who stared up at him as he wagged his tail. When the wolf moved off, the bear started to back down the trunk. But 2 ran back and the bear went up the tree again. The wolf then jumped up and put his front paws on the trunk, blocking the bear's descent. The bear tried to come down three more times, and each time the wolf leapt up and tried to bite its rear end. The bear eventually got to the ground and chased the wolf, who then turned around and chased the bear. At that point, the female ran in and both wolves went after the bear. It confronted 7 and tried to swat her with a front paw but missed. In frustration, the bear climbed another tree to wait out the wolves.

I later saw the female go to the pups with part of an elk calf in her mouth. I looked back at 2 and saw that he and the same black bear were now in a confrontation over a calf carcass. The bear grabbed the calf, ran off, then put the calf down and fed. The wolf stole the carcass back and ended up with it for good. His job would have been much harder and more dangerous if the bear had been a grizzly rather than a black. 2 later went to the pups and regurgitated parts of the carcass to them. On another occasion, I saw one of the yearlings regurgitate meat for the pups. If the yearlings did that on a regular basis, it would be a big help to the parents.

LINDA AND HER crew studied the Leopold, Rose Creek, and Druid dens in 1997. The next year they monitored the same three packs, along with the Chief Joseph wolves. In her thesis, Linda writes extensively about members of the pack, other than the parents, who help to care for pups at dens and rendezvous sites. These could be older siblings, aunts, uncles, grandparents, or unrelated wolves who have been allowed to join the pack. That communal effort by a multigenerational extended family is known as cooperative breeding and, according to researchers Dieter Lukas and Tim Clutton-Brock at the University of Cambridge in England, it occurs in less than 1 percent of mammal species.

Linda noted that yearlings and young adults "are extremely attracted to pups." They feed them, play with them, lick them affectionately, and help keep them safe from predators. Linda saw yearlings nudging pups toward the den and carrying them inside. All that interaction integrates and socializes the pups into the wolf family structure.

Linda also documented communal nursing, which takes place when there are at least two mothers in the pack. Each female nurses whichever pups come to her, both her own and ones born to another mother. In the two years of her study, Linda did not see any non-mother wolves nurse pups, but that behavior has been documented in dogs and captive wolves. Tracy Ane Brooks, a volunteer on Linda's study who cared for captive wolves at Mission: Wolf in Colorado for thirty years, told me that one spring two adult sisters appeared to be pregnant, despite never having been with a male in the breeding season. When Tracy examined the sisters, both had distended nipples and were producing milk. They were exhibiting signs of a pseudo or false pregnancy. If a truly pregnant female had been in the same enclosure, both sisters could have helped nurse the litter. That implies that the same thing could happen with wild wolves.

There can also be communal denning. In those cases, two or more mothers and their pups share a den. In 1997, Druid sisters 41 and 42 used the same den. In the Rose Creek pack, 9 and her daughter would do the same thing in 1998. There were three non-parental helpers at the Druid den in 1997: grandmother 39; 40, who was aunt to the pups; and 31, the young male who joined the pack. He was the presumed uncle to the litter. All five pups survived, and the following spring, when they were a year old, they were ready to help with the new set of pups.

Linda's faculty adviser, Jane Packard, had previously divided wolf pup development into three stages: Milk Dependent (one to five weeks), Transition (five to ten weeks as the pups are weaned from milk to solid food), and Milk Independent (eleven weeks and beyond when the pups eat only solid food). The mother relies on other pack members to bring her food during the first stage, then needs them to convey larger amounts in the second and third stages for her and the pups.

Since wolves normally reach sexual maturity at twenty-two months, helping their parents with pups when they are yearlings is perfectly timed to get young wolves prepared for adult life. Yearling male and female wolves who care for pups benefit in the long run by applying what they have learned when they later have pups of their own. In that sense they serve an apprenticeship with their parents, gaining practical experience in pup care and feeding so they know what to do when they have their first litter.

After reading Linda's thesis, I thought about my observations at wolf dens over the years. The hundreds of times I saw young adult wolves carry or regurgitate meat to the mother wolf and pups reminded me of truck drivers delivering packages of groceries from online stores to homes. Wolves invented a home-delivery system long before humans did. That system is a win-win-win situation for every member of the wolf pack. The mother benefits from getting help and food from the yearlings, the pups benefit by receiving food and protection, and those young adults profit by gaining experience in raising pups.

Most young males leave home after one year of assisting their parents with new pups. 21 helped raise pups during two denning seasons (1996 and 1997), so he was especially well trained when the time came for him to strike out on his own. Not only did he gain an extra year of experience in pup care, he also prolonged his apprenticeship with 8, and that appeared to strengthen the bond between them.

My summer job ended in the first week of September, and I had to move out of my government trailer at Tower. I rented a small log cabin in the town of Silver Gate, with a population of twenty, just outside the park's Northeast Entrance,

14

Romeo and Juliet in Yellowstone

THE DRUID AND Rose packs left the valleys in early September and went up into higher country, where most of the elk had migrated to find better feeding areas. That put the two packs out of our sight. I traveled away from the park for nine days and got back in the field on September 15. On that day I saw 8, 21, and two pups near the Crystal Creek pen, 8's original home when his family arrived from Canada and now part of the territory belonging to the Rose Creek pack. The pups and 21 gave 8 submissive greetings. After that, 8 did a raised-leg urination as a territorial marker and 21 marked over the site, making it a male-male double-scent mark. The wolves moved east, and I later got their signals in Lamar Valley, from the area south of the Institute. That put them in Druid territory.

The Druid wolves returned to Lamar on September 18, and I saw all eleven adults and pups at the Chalcedony Creek rendezvous site. 40 led them west. They traveled through the section of their territory where the Rose wolves had been three days earlier and must have gotten their scent. Likely following their trail, the Druids ended up in Rose Creek territory in the Crystal Creek pen area, where 8 and 21 had recently done raised-leg urinations. The Druid alpha pair made several male-female double-scent marks there, like a gang spray-painting their name in a rival gang's territory.

Four of the wolves—the alpha pair, young male 31, and 42—did a lot of vigorous ground scratching at one spot, leaving more proof of their foray into Rose territory. They then traveled farther west, deeper into the other pack's home. After investigating the Rose Creek den site in Little America, the Druids went back east and bedded in the Crystal Creek pen area, near 8 and 21's scent marks. In early October, the Druids made another incursion into the Rose Creek territory, visiting the den site in Little America once again and doing a lot of scent marking there.

What did the Druids think when they got the scents of the Rose Creek wolves, especially 21's, the beta male, and what did 21 think when he got the scent of the Druids? 42, the Druid beta female, was always on those exploratory trips. Her scent trail and ground scratchings would be obvious to any Rose wolf sniffing along their route. Wolves can learn a lot by investigating the scent of another wolf. When 21 sniffed the male-female double-scent marks of the Druid alpha pair, he could probably tell their rank. The beta female's squat urination would have smelled different from the alpha female's, and he likely discerned that 42 was a subordinate adult female. And I think 42 could figure out that there was a second adult male in the Rose Creek pack when she sniffed the male-male double-scent mark made by 8 and 21. After 21 and 42 repeatedly got each other's scent, perhaps each wolf began to think that this could lead to finding a mate.

If 21 and 42 both broke away from their packs and found each other, they could establish a territory of their own and start a new pack. They could breed in February and have pups in April. For 42, that would get her away from her difficult sister. But the two packs were rivals and had fought in the past. All that made me think this ongoing story was a lot like the Shakespeare play about two teenagers from enemy families who fell in love. That story did not end well for Romeo and Juliet.

What would happen if 21 left his family and traveled into the Druid territory looking for 42? I pictured the two adult Druid males attacking 21 and trying to kill him. If 42 decided she could no longer cope with her sister's aggression and headed out to search for 21, the other Rose wolves would see her as an enemy from the pack that had killed two of their family members, and they would try to kill her. Romeo was a Montague and Juliet was a Capulet. 21 was a Rose wolf and 42 was a Druid. It was a similar story. Shakespeare's play was a tragedy where both lovers died, not a romantic comedy with a guaranteed happy ending. How would the story of 21 and 42 play out?

In late fall ten Druids, all but 39, were bedded down on the south side of Lamar Valley. Something caused the big alpha male to jump up. The other adults did the same. After looking west, they ran in that direction. Then they stopped and howled as they continued to stare west. The pups whined and seemed afraid. I heard another pack howling back at them. I did a check and got signals from the Rose Creek alphas and 21. The two packs howled back and forth for some time. I drove west and spotted eleven Rose Creek wolves north of the road, three miles northwest of the Druids. They howled as they looked toward the Druids. The two rival packs were nearly evenly matched: ten to eleven.

Soon the Rose Creek wolves moved uphill to the north. But they stopped before going far and howled again, facing the Druids. After a while, the pack calmly went farther north, and I lost sight of them. I looked east and saw the Druids had also retreated. The howling contest had lasted thirty minutes, and there had been eleven separate bouts of howling between the two groups. It ended up as a draw. Apparently, neither pack wanted to risk a battle when the odds were so even. Also, each pack had pups with them. The adult wolves would know that during a fight some of their pups could be killed by the opposing side.

When the Rose Creek wolves howled at the Druids and the Druids howled back, it gave 21 and 42 another chance to be aware of each other. Breeding hormones would be peaking in a few months in both males and females. 21 would be almost three years old by then. The Wolf Project figured out that the average life span of a Yellowstone wolf is only about five to six years. 21 needed to get going on pairing off with a female and fathering pups.

AS THE SEASON progressed, I noticed increased aggression among the Druid females. Grandmother 39 often bedded down apart from the others. One day I saw a bloody bite mark on her hip that was probably inflicted by her aggressive daughter 40. An ear was also wounded. When one of the black subordinate sisters moved toward her, 39 ran off in a fearful manner. The pack had a fresh kill in the area. The old white female waited for the pack to leave the site before she fed.

In the following days, she continued to be very subordinate to the other females. Whenever 40 approached, she would go into a low crouch, tuck in her tail, and put her ears back, acting like a bullied kid who thinks she is about to be hit. One day I saw her bedded down apart from the pack at the rendezvous site. The other adults left on a hunt and passed by her without any greeting or acknowledgment. She stayed back with the pups. Later 41, the third-ranking female out of the four, came back and joined them. She still tolerated the old female.

The Druids soon left the rendezvous site at Chalcedony Creek and brought the pups five miles east to Round Prairie, a large meadow near Pebble Creek Campground. They killed a cow elk there. A grizzly showed up the next day, likely attracted by the carcass. Only 38 and 41 were with the pups at the time, and they kept the bear away from them. The grizzly took over the carcass the following day, and the wolves had to wait for it to leave before they could resume feeding.

The dominance behavior and squabbling among the three higher-ranking females kept escalating. 40 would pin 42, and she would pin 41, who often bedded down well away from her sisters. The grandmother was in the area, but mostly stayed away from the other females. Late one evening, 39 tried to approach the others, but when 40 and one of the

black sisters moved toward her, she ran off. Later the alpha
female chased 39 and caught and pinned her. I saw the
daughter biting her mother and heard the older wolf yelping
in pain.

It is hard enough for a wolf pack to support a single litter
of pups. The Druids had four adult females able to have pups
and only two adult males. There would be poor pup survival if all the females had litters the following spring. If 40
succeeded in driving off the two lowest-ranking females or
stressed them to the point where they either did not get pregnant or did not bring their embryos to term, her own pups
would have a much better chance of surviving. So far, she had
not turned on 42. By mid-October, 39 and 41 usually stayed
behind with the pups when the other adults left on a hunt.
The older female normally left the area as soon as the other
adults returned, while 41 tried to fit in as best as she could.

When one of the Druid black pups accidently got caught
in a leghold trap set by coyote researchers, a radio collar was
put around her neck, and she was designated wolf 103. She
would grow up to be the smallest female in the pack. The
pup was back in the family group the next day. We noticed
she limped on the leg caught in the trap, and when the
adults and other pups left the rendezvous site the following
day, she stayed behind and hunted voles. The next day she
was still alone. She did a lot of howling, trying to contact her
family. Later her signal indicated she went west in search of
the others. I spotted her on Jasper Bench, then saw her climb
up on a big rock, look around, and howl. She did not get an
answer. After that she walked around, trying to pick up the
scent trail of her pack.

I then saw 39 in the area. Whenever the pup howled, 39 looked around to locate her, but did not howl back. Maybe she was concerned about being attacked by 40 if she was with the pup. Moving toward the sound of the howls, the old wolf stopped on a rise and saw the pup was alone. Now that she knew there were no other adults in the area, she ran to the pup, who saw her and raced toward her. The two wolves had a friendly greeting. As the day progressed, the pup frequently howled as she tried to contact the rest of the pack, but the old wolf kept a low profile and never joined in. The two were still together at Chalcedony Creek the next day, and this time they both howled regularly. On the following day, the pup was back with the pack, but 39 left the area and was alone again.

As I continued to monitor the Druids, I noticed that 42 was normally with the other adults and pups and that her gray sister, 40, was not being overly aggressive to her. 42 played with the pups a lot. 39 continued to stay away and seemed afraid of all the females. I often saw 41 with the pups at the rendezvous site. She would run off with her tail tucked between her legs when 40 returned to the site. By early November, she was staying at least half a mile away from her domineering sister.

Despite her low status, 41 found a way she could serve her family and demonstrate her value. In the fall, as grizzlies prepare to den, their appetites greatly increase. When a wolf pack makes a kill at this time of year, one or more grizzlies nearly always show up with the intention of taking the carcass from them. Prior to the return of the wolves, Yellowstone bears had a hard time finding carcasses in the pre-denning period. In the fall, large prey animals like elk and bison are usually in prime condition and death by natural causes is not common. The kills wolves make at this time of year are perfectly timed for bears, providing abundant, highly nutritious food when they need it the most.

41 was often the pack's main grizzly deterrent. When spotting a bear, she would charge, run off when chased, then circle back to continue harassing it. She would get behind the grizzly, nip its rear end, then come back and do it again. Able to outrun any bear, the long-suffering wolf, who was so picked upon by her aggressive sister, seemed to relish those chases. It was as though she was showing off. During one interaction I witnessed, a grizzly swung a front paw at her. The wolf ducked. The paw just barely missed her lowered back.

Other wolf packs also had to contend with grizzlies. One day I saw the Rose Creek wolves near the Crystal Creek pen site. A grizzly approached the pack. The wolves ran over and surrounded the bear. It stood its ground, then spun around and swatted at a black close to its rear end. Another wolf nipped it on the behind. The bear sat down to protect that sensitive area but was bitten there again when it got up.

I got the impression that the grizzly and the Rose wolves were well acquainted with each other and had gone through this routine many times as the bear hung around waiting for the pack to make a kill. The bear eventually moved off. With the threat removed, the pups played vigorously, running around and sliding down a snowy slope. When the grizzly returned, the pups ran over and harassed it once again. One pup looked like it was dancing right in front of the bear's face as a playful taunt. The bear did not react, and the pups got bored and walked away.

Later when the pups chased a bull elk, the grizzly followed them, probably hoping it could steal the carcass if the hunt was successful. The pups noticed the bear and ran back. Two pups teased the bear from the front while another pup got behind the bear and followed it from just a few feet away. The bear mostly ignored them. 8 joined the pups and confronted the bear with his tail held high. The bear moved off and the wolves followed. Then the pack saw a herd of elk and chased them with 8 leading. The bear trailed them once again. The pups ran back and surrounded the bear. At this point, the grizzly sat up and calmly watched the pups, looking like a librarian waiting for a group of children to settle down at story time. Losing interest in the intruder, the pups ran off to catch up with the adults.

THE DRUIDS AND Rose Creek wolves continued to trade messages, sometimes by scent marking and other times by howling. I heard the Druids howling from their Chalcedony Creek rendezvous site one morning. Bob Landis told me later that the Rose Creek wolves had howled back from Slough Creek, about seven miles away. Some wolf researchers say wolves can hear other wolves howling from ten miles away, so the two packs must have heard each other.

I went to Slough Creek and spotted fifteen Rose Creek wolves in the area. I saw 8 and 9 walk side by side. He put his chin over her back in an affectionate manner. But then a pup ran over, squeezed in between the two adults, and ruined the moment. Later a big cluster of pups ran to 8 and mobbed him for attention. That day, November 2, was the last time I saw 21 with the Rose Creek wolves. Right after that he dispersed to find a mate.

I went out one more time, early on November 7, but I did not see any wolves. I returned to my cabin in Silver Gate, loaded my belongings in my van, and headed south to Big Bend for the winter.

15

When 21 Met 42

IN LATE NOVEMBER 1997, tragedy struck the Druid pack. By that time, the aggressive 40 had driven the two lowest-ranking females, 39 and 41, from the pack. The four remaining adults, along with the five pups, had traveled out of the park to the east. Both adult males were illegally shot there. The young male, wolf 31, soon died of his wounds, but the big alpha male, wolf 38, lingered on, unable to move because of his injuries. Doug Smith repeatedly flew over his location and even dropped meat for him, but he never ate it. After eleven days, the male died. His normal weight at that time was 125 pounds, but he was down to just 88 at the time of his death. The two adult females took the pups back to Lamar Valley. The breeding season was just over two months away, and the family desperately needed an adult male to join the pack in the alpha position.

Then 21 showed up. If his original plan had been to draw off 42 and start a new pack with her, the death of both Druid males changed everything. 21 was an ideal candidate for the vacant alpha male position. But if he joined the pack, he would be in a relationship with two sisters, rather than with just 42. By that time, I knew 21 and both females very well. 21 and 42 had similar easygoing personalities, whereas 40's was violent and domineering. Her temperament would not be compatible with his.

The following account of what took place on December 8 is taken from a research paper written by Dan Stahler, Doug Smith, and Bob Landis. 21 was first seen traveling with the Druid grandmother, who was still living in the area as a lone wolf. Realizing they were approaching the Druid wolves, 39 split off from 21. He continued toward the pack alone. The Druids saw him, and 40 and 42 charged at him. Three of the five pups went with them. 21 ran from the five wolves, but not in a fearful manner. The Druids soon came to a halt and 21 stopped as well. The pack howled, and he howled back.

The Druids turned around and walked away. 21 followed them. The pack chased him again but stopped when he stopped. 21 wagged his tail as he looked back at them. The Druids howled and 21 howled as well. Based on their howls, he would have known that there were no adult males in the group, just adult females and pups. He was now probably realizing this was an opportunity to join the pack as the new alpha male.

By that time, four of the Druid pups had been radio collared. Gray female pup 106 took the initiative and moved toward 21. He trotted in her direction wagging his tail, indicating a friendly intent. When they were a few yards apart, the pup did play bows to him, signaling that she also was friendly. After a few more moments of wagging his tail, 21 ran off in a chase invitation posture, and the pup followed. But she soon turned back, uncertain of the situation, and lay down.

21 moved toward the bedded pup. She jumped up and went to him. 40 ran to him at that same moment. She and the gray pup ended up on either side of him. As 21 wagged his tail, the alpha female did three play bows to him. After more friendly interactions, he followed the two females, and all three went into a forest. 21 and 40 later came back into sight. Both were wagging their tails at each other. She did more play bows to him. Then he ran off. She turned back and joined the pup.

Soon 21 returned. 42 had watched him earlier, but the two had not yet met up. By that time Bob Landis was filming the interactions, and he loaned me a copy of his silent footage. The sequence starts with 42 looking intently toward 21. Her ears are pointed forward and she is wagging her tail. Movements of her mouth indicate she is vocalizing. Then she does several romping and prancing motions toward him. She jumps up and down, like a dog happy to see a canine friend approaching.

Bob cuts to 21 at that point. He is staring at 42 in a relaxed manner. They are just a few lengths apart. She advances in small jumps. He wags his tail. Their faces are now just inches apart. He lifts a front paw and gently puts it on her shoulder. They are positioned cheek to cheek. She does little jump feints at him, lunging and snapping in the air like she is teasing him. Or maybe she is flirting.

He again puts a paw over her shoulder, in what looks like a tender, affectionate gesture. She sets her chin on his shoulder several times, and he does the same to her. Then she playfully hits him on the back with a front paw. After that she bumps her face against his. 42 backs off, runs at him, and bounces against his chest. He jumps up on her back and again puts a paw over her shoulder. A pup runs up to 21, but he continues to give 42 all his attention. He licks the fur on her shoulder and puts his chin over her back three more times. The interactions between the two are more intimate and emotional than anything that took place between him and 40.

40 now runs over and playfully jumps up on him. Both sisters stand on either side of 21, competing for his attention. The alpha female does play bows to him. Then all three romp off playfully, looking like young pups. The five pups join them. 21 is now with the entire pack of seven. All the Druids mob him like they are greeting the pack's alpha male as he returns home from a hunt.

At that point, 21 became the Druid alpha male. He had started the day as a wandering prince, and now he was the Druid king. But the Druid queen was the domineering 40, and that was going to cause him a lot of trouble. As 8 had done with 21 and his seven siblings in the fall of 1995, 21 adopted the five pups born to the Druids' original alpha male. Those pups were about the same age 21 had been when 8 had adopted him. From that moment on, anyone seeing 21 and the Druid wolves would have no reason to suspect he was not the pups' biological father.

I was especially interested in the similarities in the accounts of 8 and 21 joining a new pack. 8 first met up with the two pups outside the Rose Creek pen, and 21 first interacted with the gray Druid pup. Both adult males began bonding with pups, then met up with the adult females in the families. In human terms, it would be like a single man meeting and befriending a group of young brothers and sisters whose father had died, then marrying their widowed mother.

There were some differences between the two cases. 8 was only eighteen months old, sixteen years in human terms, when he joined the Rose Creek pack and had never raised pups or even seen one other than his siblings. 21 was thirty months old, twenty-four in human years, when he became a Druid. He had apprenticed with 8 for two years and helped him raise pups in each of those years (1996 and 1997). That meant 21 was far better prepared to be an alpha male and adoptive father than 8 had been.

Another difference was the issue of internal pack relationships. During 21's time with 8, the Rose Creek pack was a well-functioning organization in which the adult males and females got along well. But the Druid pack was a dysfunctional family due to the aggressiveness of 40 to the other females.

There was an additional downside to 21's new membership in the Druid pack: he inherited two ongoing feuds. The Druids had killed the original Crystal Creek alpha male, wolf 4, in the spring of 1996, injured the alpha female, wolf 5, and probably killed her pups. The two surviving members of the pack, the older female and the young male, wolf 6, had abandoned their territory in Lamar Valley and settled in Pelican Valley. The pair had six pups in the spring of 1997, and they all survived, upping the pack count to eight.

It had been assumed that the two wolves were mother and son and would not mate with each other, but genetic testing later revealed they were aunt and nephew. If the Crystal Creek wolves had good pup survival for another year or two, they could come back north and try to reclaim their original territory. That would involve a rematch against the Druid wolves, and 21, as the primary defender of the pack, would have to fight them. That specifically meant fighting their alpha male, 8's largest brother, who at 141 pounds was likely the biggest wolf in Yellowstone. At that time, 21 was about 120 pounds.

There was also a rival pack to the west. The Druids had killed two of 21's siblings in the Rose Creek pack: his brother on that day in 1996 when 8 had defeated 38, then his sister in the spring of 1997, whose four pups had starved after her death. Those four pups were the sons and daughters of 8. The Rose Creek wolves, particularly 8, had good reason to hold a grudge against their neighbors to the east.

How would 8 and 9 view their son's joining an enemy pack? From their point of view, he could be considered a traitor. If the Rose Creek and Druid wolves had another battle, what would happen if 8 and 21 came face to face? The two alpha males would not back down from protecting their families. If they fought, it would be father versus his adopted son.

WHEN I GOT back to Big Bend in late fall 1997, I thought back over the six months I had spent in Yellowstone that year. I had gone out looking for wolves on 170 days and had seen them on 149 or 88 percent of those days. My wolf sightings totaled 1,462, way above what I had seen in 1995 and 1996 and about twice the total sightings I had in Denali over fifteen summers. The key to seeing all those wolves and observing a wide range of their behaviors was getting out well before sunrise every day, regardless of the weather or how tired I was. I often recalled the saying "Eighty percent of success is just showing up." That's what I did. I showed up early every day.

And I was not the only one seeing so many wolves. The increased visibility of the packs enabled me to help tens of thousands of park visitors see their first wild wolves, something I never tired of doing and regarded as a great privilege. There would soon be an economic survey of park visitors done by Professor John Duffield of the University of Montana that asked people why they had come to Yellowstone. The survey found that having wolves in the park generated $35.5 million annually in tourist revenue in local communities. Seeing wolves in Yellowstone had become such a draw that many local people started wildlife tour companies that specialized in taking visitors out to watch wolves. Those businesses were creating a lot of jobs in nearby towns.

16

A New Era
for the Druids

I N THE SPRING of 1998, I arranged with Doug Smith to switch over from my job as wolf interpreter to work directly for the Wolf Project. My new job would be split between documenting wolf behavior and reaching out to the public. I had already spent so many hours as a Wolf Project volunteer that it seemed a natural transition for me. Although I would still be a Park Service employee and would be wearing a ranger uniform, no government quarters came with that position. I rented the Silver Gate cabin again. This would be my twentieth summer of watching wolves: fifteen in Denali, one in Glacier, and four in Yellowstone.

As I drove back up to the park, I thought about one wolf I never would see again. Grandmother 39 left Lamar Valley shortly after 21 joined the Druids and was later seen with wolf 52, one of 8's sons who had dispersed from the Rose Creek pack. In early March 1998, she had been spotted east of the park, traveling through a ranch. 39 ignored the livestock there but was shot and killed anyway. The man who killed her said he mistook her for a coyote. At the time she was big for a female wolf, about 125 pounds. Most local coyotes range between 25 and 35 pounds. The shooter reported the incident, pleaded guilty to killing her, and was fined $500.

After 39's death, 52 ended up with her daughter 41, who had also been harassed into leaving the Druids. The pair got together in March after the breeding season and settled into the Sunlight Basin area, just east of Yellowstone's border. They became known as the Sunlight pack. They led that pack for many years and raised numerous pups there. All those pups were the grandchildren of 39.

There had been a few other developments in the Druid pack while I had been away. Jim Halfpenny reported that 21 had mated with 40 and 42, in February. The pack was once again based at their den site north of the Footbridge and Hitching Post parking lots. All five Druid pups from 1997 had survived the winter and were now yearlings. That added up to eight wolves in the pack. The thick forest surrounding the den site had so far blocked sightings of the new pups.

One of the male yearlings, wolf 104, had killed a bison calf by himself in early spring, only the fourth known time a bison had been taken by wolves since the reintroduction began. I had been impressed by him the previous year when he was a pup, and his solo bison kill indicated he was becoming a proficient hunter. He must have spotted that bison calf, perhaps away from its mother, and tried to get it. Unlike the wolves captured in Alberta in 1995, the wolves from British Columbia had bison in their area. The older Druid wolves must have taught 104, who was born in Yellowstone, that they were a potential prey animal.

I got news of the other packs, as well. In the Rose Creek pack, 8 had again bred both 9 and her daughter, wolf 18, and the two females were sharing the pack's original den site near Mom's Ridge. Eleven pups were spotted there. Linda's den study crew felt that five were born to 9 and the other six to her daughter. Wolf 8 had sired six litters in just three years: one in 1996, three in 1997, and two more in 1998. The pups born in those litters added up to thirty-six. The Rose Creek pack now had fourteen adults and yearlings. One of the eleven new pups died young, so the pack numbered twenty-four.

In Pelican Valley, the Crystal Creek alphas had produced nine pups in April. With eight adults and yearlings in the pack, the group totaled seventeen. That meant the pack was twice the size of the Druid group. Several Crystal wolves had been radio collared earlier in the year, including a nine-month-old male pup. He weighed 115 pounds, huge for his age, but understandable since his father, wolf 6, one of 8's black brothers, was probably the biggest wolf in the park. Also the pack was very successful in hunting, and all the members were eating well.

There were now nine adults and yearlings in the Leopold pack. The alpha female had five new pups, bringing the pack count to fourteen. Farther to the west, the Chief Joseph pack had six adults and yearlings and seven new pups, for a total membership of thirteen.

Three of the major packs in the northern half of Yellowstone (Rose Creek, Crystal Creek, and Leopold) now had alpha males who were brothers from the Crystal Creek pack. A fourth pack, the Druids, had 21, who had been raised and trained by one of those brothers, wolf 8.

I ARRIVED BACK in the park late on May 5, 1998, moved into the Silver Gate cabin, and went out early the next morning to check on the Druids. As my new job had not yet started, I was on my own time.

No wolves were visible when I looked at the pack's den area from the road, so I hiked up Dead Puppy Hill. I saw two yearlings to the west and later spotted another yearling south of the Yellowstone Institute. The following morning, May 7, I found 21 on a new elk carcass near the Chalcedony Creek rendezvous site. He had probably killed the cow by himself during the night. Doug Smith speaks of May as being a time of heightened vulnerability for elk, for they have not yet shaken off the effects of the long Yellowstone winter. As I watched elk during that month, I could see that they had lost significant weight due to their low-quality winter diet.

My first sighting of 21 with 40 and 42 was on May 9. The trio was traveling south of the den area. The wolves ignored a herd of elk, a sign they were heading to a new carcass. The alpha pair double-scent marked a site, an indication of their status. Then I lost sight of them. I saw 21 coming back from that direction three hours later. He crossed the road and headed up to the den, likely to bring food to the new pups.

On an early May morning, I saw two yearlings (black female 105 and gray male 107) greeting 21 with the same excitement and passion 21 and his siblings had displayed when they greeted 8 after he adopted them. Just moments before, 107 had been watching a large herd of elk cows and sniffing the air as though he was evaluating their scent. I thought about how many dogs naturally detect health problems such as cancer in their human companions, while other dogs learn how to do so when trained. That skill comes from their wolf ancestors. That yearling had probably learned to sniff the air to analyze the various scents for hints of infections, disease, and injury. If he detected something, he could circle the elk herd, try to determine which animal was emitting that scent, approach it, and test its vigor. It was an efficient method of sorting out healthy animals from sick ones, a way to work smarter, rather than just harder.

That behavior would be like a carnival barker in past times looking to identify a mark, a gullible country boy he could talk into paying to see a fake two-headed chicken. Or an expert poker player looking for a tell in another player, some facial expression that gave away whether he was bluffing. Both the barker and the poker player would be on the look-out for vulnerable individuals. With wolves, that translates to sniffing for abnormal scents in a prey animal that would indicate a possible weakness and give the wolves an advantage.

Wolves also learn to recognize patterns: the sight of a limping prey animal or a lone elk or bison apart from the herd. A limping animal is almost always worth pursuing, and a lone animal can be approached and tested. If a solitary animal defiantly stands its ground when a wolf approaches, it usually means it is strong, healthy, and capable of fighting off the wolf. But if it runs off when it sees a wolf, that means it likely feels weak and vulnerable. An experienced wolf would leave the first animal alone and chase the second.

Well after the wolves had consumed it, Jason Wilson and I hiked out to the elk carcass we had seen 21 feeding at on May 7. We found that the cow's molars were worn down to the gumline, proof she was very old and, like a person with worn-out teeth in the pre-dentist era, unable to eat efficiently. She was likely in very poor shape when 21 killed her. She probably had infected gums as well as worn-out teeth. Perhaps he got the scent of that infection, approached her, and found her easy to kill due to her weakened condition. I later contacted Rolf Peterson, the Isle Royale wolf researcher, on that issue, and he told me that he can get the scent of infections when he examines jawbones from dead moose. If a human can do that, a wolf with a far superior sense of smell would be able to detect an infection from much farther away.

In mid-May, I saw distended nipples under 40, signifying she was nursing. Much of the fur on her belly was missing, another sign she had pups. This would be the first litter both for her and for 21. It was a new era in Yellowstone: 21 was the Druid alpha male. And I had the good luck to be there to witness how his presence would change the dynamics of the pack, especially the Druids' relationship with their neighbors, the Rose Creek wolves.

MY NEW POSITION with the Wolf Project started on May 18. I went out early every morning, looked for wolves, and monitored and recorded their behavior when I found them. Linda Thurston was still doing her den site study, and I often helped doing observation shifts. Twice a week, rotating pairs of Wolf Project staff and volunteers monitored four dens

(Druid, Rose Creek, Leopold, and Chief Joseph) for twenty-four hours, split into three eight-hour shifts.

We recorded the direction of all signals from the wolves' tracking collars at thirty-minute intervals and noted the behavior and location of any wolves in sight. During the day, when monitoring the Druid den, we hiked up Dead Puppy Hill and watched the den forest and surrounding areas from there. When I had the night shift, I parked my van in the Footbridge lot and did just the thirty-minute signal checks. I slept between checks and set an alarm clock to wake me up in time for the next one.

Another Wolf Project research study at that time was called Food for the Masses. It was part of a PhD project conducted by Chris Wilmers, and it documented which animals scavenged at kills made by wolves. When a carcass was visible, we counted all the animals, including birds, within five hundred meters of the site every fifteen minutes.

The Wolf Project was part of the Yellowstone Center for Resources (YCR). YCR employees, mostly biologists, some with PhD degrees, did research on the park's wildlife, plant communities, and geothermal features. Their main task was resource monitoring and management.

Both the Wolf Project's lead biologist, Doug Smith, and I felt we had an obligation to share what we were learning about the park wolves with visitors. Because of my years as a naturalist, that was already the normal thing for me to do. I continued to invite people to see wolves through my spotting scope and talked to them about the reintroduction program. I no longer had scheduled evening programs or nature walks. Instead, I sought out kids on school field trips, college classes visiting the park, wildlife tour groups, and regular park visitors. After helping people see wolves, I gave them impromptu roadside talks. In some years, I did about two hundred of those talks, far more than I had done when I had been a park naturalist.

In my new job with the Wolf Project, I always tried to keep the Yellowstone Mission Statement in mind. For me, the most important parts of that document were

> Yellowstone is home of the grizzly bear and wolf and free-ranging herds of bison and elk.... The National Park Service preserves, unimpaired, these and other natural and cultural resources and values for the enjoyment, education, and inspiration of this and future generations.

The key word in that mission statement is *for*. We preserve the wildlife and natural features of the park for a reason. We do it for the enjoyment, education, and inspiration of park visitors, current and future.

The enjoyment and education components of that mission were easy to accomplish. Everyone who saw wolves through my scope greatly enjoyed the experience, and all were eager to learn about them. The inspiration part came in when I talked about early rangers killing all of Yellowstone's original wolves and how we rectified that mistake by mounting a reintroduction program. It was an inspiring story. Telling people about wolf 8 going from being a picked-on pup to an alpha male was also inspirational. That was my mission and I stayed on it.

Another key word in the statement was *unimpaired*. That meant animals such as wolves and grizzlies should be allowed to live their lives unimpaired by people visiting

Yellowstone. Rangers needed to manage situations when people approached animals or blocked them when they needed to cross the road or return to a feeding area. It also meant we had to prevent habituation of bears and wolves to humans so if they left the protection of the park, they would not think it was safe to be near people, some of whom might want to kill them.

THE DRUIDS WERE the primary focus of my observations in the spring of 1998. One morning when I was monitoring the den site, I watched a play session between two female yearlings: 105 and 103, her smaller sister. 103 picked up a stick and romped around in front of her sister, like she was daring her to chase her. After a lengthy pursuit, both paused to rest, then the bigger sister did a play bow to get the game going again. Both females ran in circles as they played. When 105 lagged behind, 103 stopped, threw the stick in the air, and caught it. She frolicked around with it like a golden retriever waiting for its owner to resume a game of fetch. When that did not work, she dropped the stick right in front of her playmate, then snatched it up and ran off again when her sister took the bait and lunged for it. The chase continued until 103 intentionally dropped the stick. Now it was 105's turn to grab it and run.

Once the sisters got bored with that game, they wrestled each other. I got the sense 105 was letting her little sister pin her. After seeing that behavior and other examples of pups and yearlings playing and bigger wolves letting smaller ones win, I came to think that larger siblings will sometimes pretend to lose wrestling matches to keep play sessions going.

Wolf 9 (*right*) and her daughter, wolf 7 (*left*), in the Rose Creek acclimation pen in early 1995, the first year of the Yellowstone reintroduction project. NPS/Jim Peaco

Lone wolf 10 was added to make a functional pack. After the pack was released, wolf 9 had eight pups fathered by 10. NPS/Jim Peaco

Veterinarian Mark Johnson with the last pup rescued from 9's den, thought to be 21, after 10 was shot. The family was temporarily returned to their pen. NPS/Douglas W. Smith

Wolf 21 at six months being released from the pen. He grew up to be a powerful male like his father and became famous around the world. NPS/Barry O'Neill

Wolf 8 (*far left*) next to Crystal Creek alphas, 4 and 5. Far right is one of 8's three larger brothers, all of whom bullied him in the pen. NPS/Jim Peaco

Later in 1995, yearling 8 joined the Rose Creek wolves as their alpha male. This 1999 photo of the pack shows 8 and 9 with raised tails. NPS/Douglas W. Smith

Me leading a Park Service hike to the Crystal Creek acclimation pen in 1996. I had up to 165 park visitors join me on those hikes. NPS/David Gray

Wolf watchers at Slough Creek. This is where, to protect his family, 8 fought a much larger and stronger alpha male who had killed his father. Kathie Lynch

Grizzlies often steal wolf kills. To lure a bear away, a wolf may bite it on the rear end. When the bear gives chase, the rest of the pack can feed. NPS/ Jim Peaco

Coyote and bison. Bison will charge in to help herd members if wolves attack. A big bull can be twenty times larger than a wolf. NPS/Jim Peaco

21 howls as he approaches the Druids. Their alpha male has been shot and 21 is trying to join the pack as their new alpha. Video still/Bob Landis

Druid alpha female 40 cautiously approaches 21 to evaluate if he would make a good alpha male. Video still/Bob Landis

The Druids welcome the newcomer into their pack. When 42 meets 21 (*top*), a close, long-term relationship begins. Video stills/Bob Landis

Identification photo of Druid pup after being radio collared. She is recovering from the tranquilizing drug. NPS/Douglas W. Smith

Elk are the main prey for wolves in Yellowstone. The majority of elk hunts end in failure for the wolves. NPS/Jim Peaco

Otherwise, smaller wolves would eventually avoid playing with siblings that always beat them.

When I worked in Glacier, I had a friend named Bill who owned a wolf-dog mix called Kintla. He was a big, tough-looking animal, and most people were initially afraid of him. One day I was in Bill's house with Kintla, and we began a game of chase around the dining-room table. Kintla chased me, but he deliberately ran slower than normal so that he never caught me. I stopped and looked back at him. He also stopped and studied me intently, then turned around and ran the other way. I took that as an invitation to chase him and went after him.

As I ran around that table, Kintla repeatedly looked over his shoulder at me, making sure I was still chasing him. He ran at a pace that allowed to me keep up. We both soon stopped and looked at each other, then it was his turn to chase me again. I knew he could catch and pin me any moment he chose. Kintla knew that as well, but he wanted to play and pretended to be afraid of me so he could prolong the fun. Those minutes playing with Kintla were the closest I ever came to experiencing what it might be like to be a wolf.

When I later thought about the moments when Kintla ran from me, allowing me to chase him, it reminded me of something from my childhood. My brother, Alan, was six years older than me, a big difference when you are six years old. I remember a game he invented that was great fun. Back in those days whole chickens bought in grocery stores came with the feet and claws attached. Alan pretended to be ter-rified of those feet. Each time our mother returned with a chicken, I would secretly go to the refrigerator, pull off a leg, hold it so the foot was sticking out, then sneak up and charge at my brother. Seeing the chicken foot about to grab him, like in some horror movie, he ran off, acting scared to death. I did that repeatedly to him and it always worked. But I knew it was really just a game, and he was pretending to be scared.

I thought about how the principle of a bigger wolf letting a smaller one win might apply to 21. He was so big as a pup and yearling that, like 105, he probably figured out that he needed to let his smaller brothers and sisters win their share of chases and wrestling contests to keep the play going. I decided to look for opportunities to see 21 playing with the Druid yearlings and adult females to see if he behaved that way with them.

THAT SPRING, THE Druids were having a lot of trouble getting food to the pups because their den was on the north side of the park road, and their main hunting area was to the south. People parked in the nearby lots would see the wolves approaching the road, run to their cars, and drive down to the likely crossing spot to take photos. The wolves would back off, circle around, and try another crossing point, only to get blocked there as well. The law-enforcement rangers put up No Stopping signs at the most common crossing sites near the den, and I was given a big red stop sign so that I could work as a wolf-crossing guard if necessary. 42 was the savviest Druid wolf regarding road crossings. She would run to the road, slow down, look both ways, then, if no cars were heading toward her, race across to the other side.

Every day that I observed the wolves, I learned more about them. In late May, I was on Dead Puppy Hill watching the den site to the north when I spotted 40 alone down by the creek below me. I lost her on the west side of the hill. An hour later, she suddenly appeared just downhill from my position. I was already sitting and crouched down farther, hoping she would not notice me. She looked my way, then walked toward me, sniffing the ground. I had the impression that she was not seeing me. When she crossed the route I had taken uphill and got my scent, she instantly ran back the way she came.

What I concluded from that incident was that wolves do not do well at seeing details, especially if an animal or person is motionless. The wolf had looked in my direction but seemed unsure of who or what was there. Then she had walked toward me with her nose to the ground, picked up my scent, and run off. I later learned that wolves are very good at detecting motion, even at great distances. Often, I would see a wolf traveling through Lamar Valley pause and stare intently in some direction, then go that way and chase a far-off elk I had been totally unaware of.

NOW THAT 39 and 41 were no longer with the pack, I wanted to see how the domineering alpha female treated her sister 42. On June 2, 42 was bedded apart from the others. The alpha pair and three of the yearlings ran toward her with 40 in the lead. 42 went into a submissive posture, then rolled onto her back. As soon as the alpha female reached her sister, she bit her hard several times for no apparent reason. After 40 left, 42 went to 21, and the two had a friendly greeting.

I noticed that 42 often went to 21 after aggressive treatment from her sister. I saw some film footage taken by Jim

Halfpenny around that time when 42 ran off to 21 after 40 had attacked her. The alpha female raced after her, then stopped and evaluated the situation. 21 just stood there, next to 42, in a neutral but confident posture. I had the impression that 40 did not know how to react. Whatever she was thinking, she walked off, leaving her sister alone.

Other times, 21 took the initiative to show his support for 42. Later that year, when 42 was standing off nervously by herself, wary of getting close to the alpha female, 21 left the other wolves, walked over, and stood next to her as the rest of the pack socialized. As he had with the sick pup in the spring of 1997, I think he noticed she was in distress and went over to be with her.

21 took great care to attend to the needs of his new pack. He was usually the first to bring meat from a kill back to the pups. Early one morning, the pack pulled down an adult elk at the Chalcedony Creek rendezvous site. 21 went back to the den two hours later to feed the pups while the other adults stayed at the site and slept. When they returned to the carcass in the evening, 21 joined them, then, after feeding, he trotted right back to the den. 40 seemed less concerned than 21 about bringing food to the pups.

One evening in mid-June, I heard the Druid wolves howling from their den forest. In addition to the low-pitched calls from the adults and yearlings, I heard higher-pitched howls that had to be coming from the pups. For the very first time, I was hearing pups sired by 21.

Because they spent a lot of time watching the Rose Creek dens in 1996 and 1997, I also talked with Wolf Project volunteers Debbie Lineweaver and Jason Wilson about their impressions of the relationship between 8 and 21. Debbie told me the two males seemed to have an understanding that they needed to cooperate to get things done for the pack, especially hunting and bringing food back to the den. She said they divided up the responsibilities of feeding and protecting the family and used the term "co-leadership." Jason told me he never saw the older wolf dominate 21 or witnessed the younger wolf challenge the wolf who had adopted him in any way. "They had an easy relationship," he told me, "with no dominance or class distinction." He added that it "was a partnership of equals."

I never saw 8 exert his dominance over 21, either. 8 had what I would call an even-tempered, confident personality. In the spring of 1997, when 21 was two years old, about twenty-two in human years, and surpassing 8 in size and strength, they functioned perfectly together. It was clear to me that 21 respected 8 as the pack's alpha male and as the wolf that had raised him, and 8 valued what 21 did for the family. Debbie called them co-leaders, and Jason said it was a partnership. I would add to their comments the idea of friendship. To me, the Rose Creek adult males were like two dogs who liked being together. Alpha male 8's easy confidence was in sharp contrast to the aggression shown by the Druid alpha female 40, who seemed insecure in her position and needlessly and repeatedly dominated her sister and the three younger females. As I watched her bully and beat them up, I wondered whether she was worried they might one day turn on her.

Despite 21's new responsibilities as the Druids' alpha male, including the need to get ever-increasing amounts of food to the growing pups, he took time out to play. I saw 40 approach him, then suddenly turn around and run away, an obvious invitation to play catch me if you can. He sped after her and they raced through scattered trees. Soon she turned around and charged at him. He saw her coming and ran away, pretending to be afraid of her. It was just like the time Kintla and I had traded off on chasing each other around the dining-room table.

As the pups matured, the Druid adults began to roam farther afield on hunts. Doug Smith did a tracking flight on June 21 and found the alpha pair and two yearlings out of the park to the east, near Crandall Creek, the area where 31 and 38 had been shot the previous fall. We were nervous about the situation, then relieved when the wolves were safely back in Lamar on June 23. On another flight three days later, the alphas were spotted twenty miles to the south in Pelican Valley, feeding on a bull elk carcass in the heart of Crystal Creek territory. 21 and 40 were there alone. If the eight adults and yearlings in the Crystal Creek pack got the scent of the carcass and found the two Druids in their homeland, they would attack. To protect 40, 21 would have to fight them, including the Crystal alpha male, wolf 6, one of the few males in the park who was bigger than he was. But luck was with them, and the next morning I got both of their signals back at the den forest.

As I continued to watch the pack, I began to pick up differences in the personalities of the yearlings. There was wolf 104, the enterprising black who had killed the bison in the spring. His gray brother, wolf 107, was bigger, but did not seem to have 104's initiative or willingness to risk injury. I once saw the larger brother chase a cow elk, but when he caught up to her, he just nipped at her hind legs. He looked like a dog that had caught up with the car he was chasing and now didn't know what to do with it. As he ran after the cow, he fell and tumbled over several times. I could not tell if the elk had kicked him or if he had tripped.

When he got up, he must have seen that the cow was limping badly, which should have given him a big advantage, but on catching up with her again, 107 stopped and just watched her run away. My impression of 104, in contrast, was that he never gave up when a good opportunity was in front of him. He would keep working a situation until he found a solution. If he spotted a cow with a bad limp, he would finish the job. DNA analysis later showed that he was 42's son. His larger brother was never collared, so we never knew if they were full or half brothers.

I continued to monitor how the five Druid females interacted with each other. I had already noticed that 42 was much more attentive to the yearlings than the alpha female was. On one earlier occasion, after 40 had aggressively pinned 41, I watched as 42 went to her black sister and began playing. It was like seeing someone going up to a bullied girl and befriending her.

One day in late June, yearlings 103 and 106 ran past 40 to greet 42. When the alpha female ran over in a dominant posture, 42 dropped to the ground and rolled on her back under her sister. 40 snapped at her and walked away. Cautiously, 42 got up in a low crouch. On seeing that, 40 ran back and snapped at her again. 42 rolled on the ground, then reached up and licked 40's face to appease her and demonstrate subordinate status. It seemed to help, for the alpha female allowed her to get up. I later wondered if 40 went after her sister because the young females had rushed to greet 42 rather than her.

Then I saw 21 do something that enhanced my already high opinion of him: he went to 40 and initiated a play session with her, the way a human father might play with a child that was acting out. After lightening the mood by distracting her with play, 21 ran circles around the other wolves and did play bows to them. 40 and the yearlings were soon chasing him. Then 21 turned around, and he and the yearlings chased her. After fleeing a short distance, she reversed directions and chased him once more. I saw 21 running sideways as he looked back at her. Suddenly, he dropped down in tall grass, set up an ambush, and leaped up and chased her when 40 ran in. Then he went to the yearlings and played with them.

All that time 42 stood apart from the others, as though she was unsure what would happen if she joined in and encountered her aggressive sister. I noticed all the other wolves were now chasing 40 and she was running sideways, as 21 had done, so she could look back at them in a manner that dared them to try to catch her. 21 caught up with her and the pair played together. They stood up against each other, chest to chest, and sparred with their jaws and paws. Then she chased 21 and ran circles around him when he fell, probably deliberately. When he got up, he chased her.

I looked around for 42 and saw she was now playing with the yearlings. The pack was back to normal, with all the tension dissipated, thanks to 21. He was a peacemaker who used play to restore feelings of goodwill in his family. During episodes like that, 21 acted not as the big-shot alpha male, but as the pack's court jester. He might have been the biggest and toughest male in the neighborhood, but he had no problem playing the fool.

A lot had already happened with the pack that day, but there was one more incident to come. For 42, it was a real-life example of the saying "Every dog has its day." I saw the wolves chasing a pronghorn fawn. It disappeared suddenly in a patch of thick sage, probably because it had collapsed from exhaustion. The pack sniffed around the brush, searching for it. A moment later, I saw 42 running off with the dead fawn in her mouth. 40 wagged her tail at her sister, as though she now wanted to be friends and get a share of the fawn. But 42 ignored her, ran farther away, then bedded down. She kept the fawn to herself and ate it. 40 respected her sister's right to the food and did not bother her.

From that incident I learned that low-ranking pack members have a right of ownership to food they possess, even if a higher-ranking wolf might want to take it from them. It was like seeing a big dog respecting a much smaller dog's right to the food in its bowl. 21 also killed a pronghorn fawn, but he shared it with some of the yearlings, something he regularly did. I remembered the time at Slough Creek when 8 had killed an elk calf and shared it with three of his adopted yearlings. This was another case of 21 modeling his behavior after 8.

That day was an exhausting one for me. I had gotten up at 4:00 a.m. and found the pack at 5:26 a.m. They had been in view continuously until 9:22 p.m. that night, nearly sixteen hours. But it was worth it, for I had seen some astonishing wolf behavior and helped hundreds of people see wolves.

I NEVER SAW an alpha male wolf who enjoyed playing as much as 21 did. In early July, I spotted him and 103, south of the road. They were about five hundred yards apart, out of sight to each other. He was digging at a hole, possibly a coyote burrow. She howled. 21 looked in her direction and howled back. She answered right away and ran toward him. He was still digging at the site when she arrived, but the young wolf wanted to play, so she jumped on his back and bounced up and down on him with her front paws. He ignored her at first, then turned and wrestled with her. The much-smaller female pinned him. Since he outweighed her by 40 to 50 pounds, he must have allowed her to toss him on the ground. After more play, he got up and resumed digging. She bedded down and watched as he dug a hole deep enough for him to disappear into. Then 21 gave up on the excavation and they both walked off.

The little female soon turned to him, leaped in the air, twisted around, and landed by him, inviting him to play. The big alpha male looked at her, then ran off. She chased him, caught up, and knocked him down, something he must have let her do. She then climbed on top of him, in the position of the winner of a fight. He wriggled out from under her and ran away. She pursued him. Both wolves tripped and fell. 21 got up first and sped from her again, like a low-ranking wolf from a domineering alpha. They fell again. She jumped up first and stood over him once more in the dominant position.

He got up and they wrestled. 21 allowed her to throw him to the ground. She lay beside him, and they both hit each other playfully with their front paws. Then the big male jumped up and ran off with the low tail of a submissive wolf. She caught up with him and he fell, pretending she had once again out-wrestled him.

The play session lasted thirty-five minutes and during that time 21 acted not as an alpha male, but more like a yearling or even a pup. He pretended to be a much younger and lower-ranking wolf, one that the smallest female yearling could chase and catch, then pin.

I thought back to my observations of 38 in 1997, the year the five yearlings were pups. I could not recall any sightings of him playing with them like 21 did this year. The big male had done a good job of feeding and protecting the pups, but I never saw him have the playful interactions 21 frequently had with them. The irony here was that 38 was their biological father, whereas 21 was their adopted father. It seemed to me that 21 had a better emotional connection with the family. In the end, I decided that the two males just had different per-sonalities. 38 was more aloof in his interactions, while 21 was more interactive and playful.

Three days later, I went up Dead Puppy Hill and saw 21's pups at the Druid den for the first time. There were two, a black and a gray, in front of the den forest. They would be nine or ten weeks old. We knew that 21 had bred 40 and 42 the previous winter. Since Yellowstone wolf litters aver-age four to five pups, we hoped the Druids would have a lot of pups that year, but we never saw more than those two. I heard that 42 had stayed close to what might have been a den site on a ridge just east of the Yellowstone Institute for a while, but then she abandoned the area and was based at the main den from then on. Had she given birth to pups at the first site and lost them? Or had she never been pregnant at all? I put the questions aside for now, thinking I would probably never know.

One day 103 and her sister 105 seemed to be on baby-sitting duty at the den. 103 was looking toward the nearby road, as though she was watching for anything dangerous that might harm the pups. Both pups came over to her. The black pup bedded down under her, and the lighter one licked her face. Wagging her tail at them, she suddenly sped off. They instantly understood the game and chased after her. As 21 had done with her a few days earlier, 103 pretended to flee from the pups. In contrast, 105 avoided the pups and hid in tall grass when they came toward her.

The next day, the two pups spotted male yearling 104 bedded down and ran to him, hoping to play. Seeing them coming, he got up, jumped over the backs of both pups, and ran off. The pups chased him and nipped him in the rear end, as though they were pretending to attack a fleeing elk. He ran toward 106. The pups promptly forgot about him and harassed her instead. It looked to me like a deliberate strategy on his part to switch the pups' attentions over to her. She was more willing to play with them. 106 picked up a six-foot-long stick and walked away from the pups with it. The black pup ran over and grabbed the far end of the stick, and they walked along side by side, carrying the toy. Soon both pups were running around with sticks in their jaws, leaving their older sister in peace.

Later that morning, 104 got in a playful mood, did bows to the pups, and romped around with them. He chased the black pup, caught up with it, and tried to grab its back with his mouth, but the pup squirmed away and ran off. Then the play slowed down and the pups drifted away. When they were scared by a plane flying overhead, they ran back to their big brother for protection.

On July 10, the adult wolves moved the two pups across the road and through Soda Butte Creek to a new rendezvous site southeast of Soda Butte Cone. That was the area where the rangers had killed the last two original Yellowstone wolves in 1926. A pack of ten wolves was now within a few hundred yards of where that terrible incident took place. Once considered hated outlaws to be hunted down, wolves had become one of the top Yellowstone tourist attractions and responsible for big profits for local businesses catering to park visitors. Times had changed.

At the Soda Butte Cone site, I noticed the pups seemed to prefer to be near their father more than any of the other adults. They would walk alongside 21 and bed down near him. Perhaps they felt safe and secure when they were with the big black wolf.

104 continued to impress me. One day I watched as the Druids chased pronghorn fawns. The yearling went after one fawn, and the alpha pair and three other yearlings joined in. All but 104 gave up the chase when the fawn outran the pack. An adult pronghorn ran toward him, probably trying to distract him, but he ignored it and continued after the fawn. Seven minutes into the chase, lacking the stamina of an adult, the fawn collapsed from exhaustion and the yearling

got it. He had stuck with the chase when the other Druids, even 21, had given up, and he had succeeded. 104 shared the carcass with one of his sisters, perhaps imitating what he had seen 21 do.

The black pup later followed the adults west from the Soda Butte area, and the pack ended up at the Chalcedony Creek rendezvous site, which was a better central location for them. The gray pup stayed behind. I later went up Dead Puppy Hill and saw him north of the road, back in the den area. The pup howled continuously, stopping at regular intervals to listen for answering howls. I was about eight-tenths of a mile from the pup. I sneezed and he immediately looked directly at me, an impressive demonstration of a wolf's hearing ability. 106 must have heard the pup's howling, for later I saw her leading the pup to the other pack members at Chalcedony Creek.

On July 16, three hikers walked out near the Chalcedony Creek rendezvous site. I could not see any wolves in the meadow, but I could hear the pups howling from nearby trees. Then I spotted the alpha pair coming out of the forest. With 40 leading, they ran toward the people to see what was happening. As they ran, they frequently looked back in the direction of the pups and howled, probably warning them to stay put. By that time, the three hikers had moved back toward the road and were out of sight to the wolves. The two adults stopped, looked toward the road, then ran back to the trees and the pups.

We did signal checks on the alpha pair and found they left the area an hour later, likely taking the two pups with them. The next day I spotted 105 howling at the rendezvous site. She seemed to be looking for the pups and other Druid adults. On the following day, I saw 40 and 107 in nearby areas, but I did not see any pups.

For the next few days, I continued to go out early to look for the Druids and check on their signals. From July 19 through 24, I had no signals or sightings, nor did I hear any howling. It appeared that the pack had left the valley due to the disturbance at their rendezvous site. A tracking flight on July 23 found five Druids in Pelican Valley. The pups were not with them, but they had probably been stashed elsewhere with the other three adults. Lamar Valley, without the Druid wolves, seemed empty and barren as it fell temporarily silent.

18

The Chief Joseph Pack

W E WERE UNLIKELY to see any wolves for some time in Lamar. The Rose Creek wolves were also in a remote area. Linda was doing one of her den studies on the far western side of the park, in the territory of the Chief Joseph pack, and I was reassigned there for the balance of July and into August.

The Chief Joseph pack was a blended family with seven pups and four yearlings. The two wolves who later became the alpha pair had been in the same acclimation pen when they came down from British Columbia in 1996, but they had gone their separate ways after release and did not get together as a breeding pair until the summer of 1997. That was when the female reunited with the male and helped raise the four pups that he had sired with his previous alpha female, who had died after being impaled on a branch while out hunting elk. The pups' mother was a daughter of 9, so they were her grandchildren. Now, a year later, those pups were yearlings.

I got in place on a hill overlooking the pack's rendezvous site on the evening of July 24 and soon saw the alpha pair, three of the four gray yearlings, and all seven of the three-month-old gray pups. The alpha female, wolf 33, had a sleek black coat and the alpha male, wolf 34, was a big gray. The yearlings were nearly identical, but I soon distinguished differences in the markings on their gray coats.

I settled down to watch. One pup found a stick and ran around with it. The other pups chased it. Later a pup went to a yearling that was bedded down and tried to interact with him. The yearling did not want to play and gave the pup a slight nip. He apparently held back from using any real force, for the pup continued to harass him. It nipped at the male's back and pawed at his face. As I watched the pack in the coming weeks, I saw that the pups did not take growls, snaps, or threats from the yearlings seriously. They seemed to know that wolves are hardwired not to harm young pups and ignored any attempts the yearlings made to discipline them.

My days fell into a routine. I got up early in the morning, watched the pack, took a break in a nearby rented cabin when the wolves rested in the afternoon hours, then returned for an evening shift. I did that every day for the four weeks I was stationed there.

That first morning, I noticed one of the four yearlings interacted more with the pups than the other three. I saw all seven pups surround him, trying to reach up to lick his face. He hovered over them as he wagged his tail. He looked like he was enjoying their attention. Then he lowered his head and regurgitated a few pounds of meat to them. The pups gulped down the meat as he watched. He went from one pup to another, sniffing all seven in turn, like he was checking on each one. Then he regurgitated meat for them one more time.

Later he grabbed a bone and ran off with it. The pups chased him. He dropped the bone, turned around, and chased a pup, playfully nipping at its rear end. After that he played vigorously with all seven pups, sparring and wrestling with them. I saw him get behind a pup, lunge forward, goose it on the rear end with his nose, and knock it down. Seeing this made me wonder whether wolves have a sense of humor. I concluded that if this yearling did, other wolves would have it as well. I remembered hearing a definition of what's universally funny: seeing someone fall down; tragedy is when I trip and fall.

Unlike the other three yearlings, this male repeatedly sought out the pups for play sessions. In my field notes I called him the playful yearling. I saw him energetically play with four of the pups, then run over to the other three and interact with them, giving equal time to all of them.

That morning the playful yearling spent ninety-nine minutes with the pups while two other nearby yearlings had only minimal interactions with them. The pups constantly ran back to him. They liked to swarm over him when he was bedded down. He would roll over on his back and wiggle his paws in the air at them. I saw him chase a cluster of pups, leap over them, then turn around to face them. All seven pups surrounded him, and he gently nipped at one, then another.

The pups were fascinated by the radio collar around their mother's neck. I saw one pup bed down next to her and use a front paw to bat at the collar, causing it to swing back and forth. Then the pup hit her several times in the face. Another time, a pup repeatedly nipped one of her ears. On both occasions, she patiently tolerated the pups' rough treatment.

The pups learned how much fun it was to sneak up behind a sibling, grab its tail, and yank. During wrestling matches, I saw them learn to bite the fur on the back of the neck of another pup so they could twist it to the ground. These playful matches were good preparation for real fights with rival wolves when they grew up. The yearlings expanded the pups' repertoire by demonstrating moves that could help them later in life during hunts. I saw one chase a pup, reach forward, grab one of its hind legs, and pull it down. Adult wolves employ the same maneuver to jerk the elk calf they are chasing to the ground.

As the pups played, the adults needed to be vigilant in watching for threats. One evening a yearling spotted a black bear approaching the site and slowly advanced toward it. When the bear saw the wolf, it ran and climbed the nearest tree. When it later came down, the alpha male took over the job of protecting the pups and charged it. The bear raced back up the tree and stretched out on a big branch to rest. 34 bedded down nearby and waited. When the bear climbed down, he charged it once again. As the bear started climbing back up the trunk, the wolf leapt up and bit it on the bottom. Frantically trying to get out of reach, the bear climbed higher. It ended up lying down on the same branch where it had previously rested. The father wolf walked off and bedded down where he could monitor the bear. The bear was still in the tree when night fell, and I had to leave.

Since this was our first case of a female joining a pack after the death of the previous alpha female, I watched to see how 33 interacted with her adopted pups now that they were over a year old and bigger than she was. One evening a yearling went to her, rolled on his back, and gently pawed at her face, just as a pup would to its mother. He then licked her face. She tolerated his attentions good-naturedly. Her behavior was indistinguishable from that of a biological mother.

The pack's hunts were often unsuccessful, and when no elk meat was available, the pups chewed on grass stalks. After the pack left the area, I walked out and saw scats full of undigested bright green grass, a sign it had no significant nutritional value to them. I remembered seeing pups leaping up as they tried to catch flying insects. I had also watched them grabbing and eating crickets and grasshoppers. On that hike, I came across pup scat full of the hard outer shells of those insects. One even contained the head of a grasshopper. Other scat was full of elk fur and bone chips. The chips would have been from pups gnawing on old bones, and the fur, from them chewing on elk hides.

IN MID-AUGUST, THE Yellowstone Park Foundation arranged to bring Harrison Ford and his family out to the park. Since we had such a good situation at the Chief Joseph rendezvous site, I was given the assignment. The family was going to drive up from Jackson, Wyoming, in the morning, and Benj Sinclair from the Foundation and I would take them up to our observation point in the afternoon. But the son was sick that day, so Harrison stayed home with him. His wife, Melissa Mathison, screenwriter of *E.T. the Extra-Terrestrial* and *The Black Stallion*, and their daughter, Georgia, joined us and we walked up to the viewpoint. All seven pups and one of the yearlings were out, and we watched them all evening.

We went back the next morning and found the seven pups again. Fog came through the area in patches and at times blocked our view. At one point, when the fog was clearing, I noticed the pups were arranged in a circle, something I had never seen before. The center was obscured by fog. When the fog cleared, I saw a grizzly in the middle of the pups. The bear calmly looked at the seven little wolves, then walked uphill. Unafraid, the circle of pups walked along with it. Then they reorganized and trotted after the bear in single file, as though they were following a ranger on a nature walk. The grizzly looked back casually, as though having wolf pups following it was a normal part of its life.

That sighting was one of the very best of the summer. Benj and I were glad Melissa and Georgia got to see it. We drove them to the West Yellowstone airport that afternoon, and Harrison flew in with his helicopter to take them back to Jackson. I stayed in touch with Melissa, and that fall I visited her in New York City and did a wolf talk at her daughter's school.

I later learned that the tracking flight that day had found the alpha pair over thirty miles east of the rendezvous site. They were back with their pups early the next morning. I checked my records and found that the mother had been away from the pups for at least ninety-seven hours while 34 had been gone for a minimum of seventy-two, a necessity if they were having a hard time making a successful hunt.

One of our sightings in late summer was like the climactic scene in the Clint Eastwood Western *The Good, the Bad and the Ugly*. A yearling was with two pups, and all three stood in a circle staring at each other for a long time, motionless. Then each wolf slowly crouched down, as though they were getting ready to charge and pounce. At the same moment, all three ran forward. One of the pups veered off, and the yearling and other pup chased and pinned it. That was a new game: standoff.

On the morning of August 23, I saw the Chief Joseph wolves as usual at their rendezvous site. That turned out to be my last sighting of the family. Based on signal checks, they left the site later that day.

One of those four Chief Joseph yearlings would later disperse and end up near Lamar Valley, where he would start a new pack with one of 21's daughters. They would have many pups, and one daughter, officially 832F but better known as the '06 Female, would become famous worldwide. I think the male was the playful yearling for he behaved in the same manner with his own pups as he had with the pups that summer. He would be collared and assigned the number 113.

19

Family Life with the Druids

I WENT BACK TO Lamar on August 27 and found the Druids that evening. The black pup was not with them. It had not been seen since July 16, the day of the human disturbance at the rendezvous site, forty-two days earlier. We never saw that pup again. Perhaps it got separated from the adults that day and could never find its way back to them.

Two days later, in the early morning, we spotted the pack at the Chalcedony Creek rendezvous site. The three adults, five yearlings, and the gray pup were in the group of nine. 40 was in a playful mood, somewhat unusual for her, and all the other wolves joined in. Soon the alpha pair broke away from the others and played together. They reared up, bumped against each other, and struck out with their front paws. When they dropped down, 21 pretended to be afraid of 40 and ran off with her in pursuit. He even tucked his tail, acting like he was the pack's lowest-ranking female. He was playing the fool again, letting his inner pup out.

When the pack encountered a big bull bison later that day, 21 and 104 were the ones that got closest to him. The yearling darted repeatedly at the bull, jaws snapping as though he was going to bite him. The bull charged, and the young wolf easily dodged out of the way. The others could have rushed in and attacked the hindquarters when the bull charged at the yearling, but they did not take advantage of that opportunity. Once again, 104 took more initiative than the other pack members.

In late August, the Crystal Creek pack lost its alpha male, wolf 6, the young wolf that had paired up with his aunt after the Druids had killed the original alpha male. Doug Smith had done a tracking flight on August 25 and got mortality signals from him in Pelican Valley. A crew had hiked out to the scene, examined his remains, and found that he had died from puncture wounds made by an elk antler. There was a freshly killed bull elk next to him. The wolf must have fought with the bull, been mortally wounded in the encounter, but still killed him in an impressive example of courage and determination. He also had injuries that looked like a grizzly had slashed his side with its claws. That likely happened after he got gored, when he was defending the carcass from the bear. Then he died of his wounds from the antler. If an alpha male were to choose his manner of death, that would be a heroic way to go. At the time he died, wolf 6 was probably the largest and strongest wolf in the park. Of the four Crystal Creek brothers, two remained: 8 and wolf 2, the Leopold alpha male.

On September 8, the Druids were in Lamar Valley as usual, all but 104, and I did not get his signal during the next few days. The tracking flight on September 11 found him alone in Hayden Valley, a few miles west of Pelican Valley. During the next flight, five days later, he was with the Crystal Creek wolves and had apparently become their new alpha male.

104 was only seventeen months old, still a yearling, about the same age as 8 had been when he had joined the Rose Creek pack as their new alpha male. The Crystal pack had at least two male yearlings, but he had managed to walk into their territory and become the top male. Those two brothers would have been the sons of the alpha female. Since wolves do not normally breed with close relatives, she needed a new unrelated male. Her acceptance of 104 would have overcome any resistance from her sons. That was our third case of an outside male joining a pack as the new alpha after the death of the previous male. In the previous situations, the new males, 8 and 21, had adopted and raised the pups in the new packs. We expected 104 would do the same.

WITH 104'S DEPARTURE, the Druids now numbered eight: the alpha pair, 42, the four yearlings, and the gray pup. That pup was already taller than the three female yearlings and nearly the same height as 107, the male yearling. His mother was most likely 40. He undoubtedly inherited his size from 21.

I continued to study how 21 interacted with the other Druid wolves. One evening, three of the pack members were in sight: the alpha pair and 103. The yearling and 21 playfully sparred with open jaws as they wagged their tails at each other. Pretending the much-smaller female was beating him, 21 backed up and accidently stepped on the bedded 40. She jumped up and snapped at him. Rather than engaging her, he dodged the bite, moved away, and went back to playing with the yearling.

Later 21 walked off to the southeast by himself. He repeatedly stopped and looked back at 40, seemingly wanting her to follow. He howled and she howled back. Then she got up and moved off, but went east rather than directly to him. The big male looked at her, then ran to her and went her way. All that indicated that the alpha female was the true leader of the pack. If he wanted to go one way and she went in a different direction, he followed her. She was the boss.

During a sixty-five-minute period a few days later, 40 led the pack for fifty-five minutes while 21 was out in front for only ten minutes. During that time, he stayed on the route she set. When she took over the lead position again, he followed every time she veered off in a new direction, including when she reversed direction and went back the way they had just come. He looked like a married man meekly following his wife around a department store.

As the years went by and I watched many wolf packs, that pattern was always the same. Wolves live in a matriarchal society and females run the show. I recalled a story I came across while doing research for my book *A Society of Wolves*. A biologist at the Washington Park Zoo in Portland, Oregon, had signed up a thirteen-year-old volunteer to observe the zoo's wolves. He pointed out the various pack members to her and explained that the alpha male was the leader of the pack, which was the conventional understanding at the time.

When he reviewed her observation notes, he saw that she had made a major mistake. The volunteer had listed the alpha female as the pack's dominant animal, not the alpha male. Worried because she had made such a fundamental error, he went to the wolf enclosure to observe the wolves and confirm the error before speaking to her. But as he studied the pack, he realized the assumption he and other wolf biologists, all men, had made for years was wrong. The young woman, watching the wolves with an open mind, had seen the truth when so many others had not.

ONE OF 21'S primary responsibilities was protecting his family from any threats. One fall evening, I saw the alpha pair and the gray pup bedded down at the Chalcedony Creek rendezvous site. A huge bison bull came into the area, walking directly toward 40 and forcing her to get up and move off. Two more bulls joined the first one, and all three followed the alpha female. One of the bulls ran at her, and she had to dodge him. By that time, 21 and the pup were also up and trying to avoid the bulls.

Two of the bison soon lost interest and wandered off, but the original bull continued to pester the wolves. He charged the alphas. 21 was running off behind 40, when he suddenly turned and defiantly faced the 2,000-pound bull, looking like the lone man who stood up to the tanks in Tiananmen Square. The bison stopped dead in his tracks and stared at 21. The wolf stood his ground as he blocked the bull from getting closer to the female. After twenty seconds of standoff,

21 casually turned around and trotted toward her. The bull seemed to ponder the situation for a few moments, then he, too, moved off, away from the wolves.

Every wolf in the pack, not just the alphas, has to pitch in and work together to defend and feed the family. One evening I saw 42 and 103, the smallest female yearling, traveling together. 42 was lagging well behind when, suddenly, she ran west. I looked that way and saw the yearling had a holding bite on an elk calf's rear end. 42 ran in and joined her, and they worked together to pull the calf down and kill it. Other people on the scene told me both wolves had chased the calf earlier, but after it outran them, 42 had given up. Later the little yearling went after the calf by herself, closed in on it, lunged forward, and grabbed a hind leg. The calf continued running and dragged the wolf along behind her. It kicked 103 in the face with its other hind leg, shook her off, and ran away. The wolf caught up with it again and bit into its rear end. That was when 42 saw the chase, ran in, and helped the yearling finish off the calf.

The vast majority of wolf hunts end in failure. In a recent talk, Doug Smith said that the failure rate can be as high as 95 percent. I saw a good illustration of that in late October when the eight Druids tried to kill a big bull elk. The wolves surrounded the bull as he stood his ground. 21 got behind him and bit his hind leg but let go right away, probably to avoid getting kicked. The bull confidently trotted off at a slow pace, seemingly unconcerned by the presence of the wolves. The wolves walked along with him. The bull stopped, and the wolves seemed to be studying him, looking for points of attack. When he took a few steps toward the pack, they backed off. Once he continued on, they traveled with him. The bull seemed totally calm and confident, looking like a huge NFL or NBA player walking through a dangerous part of a city, sure no one would bother him.

Then 107, a wolf who had not impressed me in the past, ran in and nipped the bull on the rear end. The yearling ran off right away but immediately came back and made a second bite at the same spot. The bull did not react. It was like an insignificant insect had bitten him. Demonstrating his indifference, the elk raised a hind leg and casually scratched his head with a hoof. That moment when the bull was off balance would have been the perfect time for the eight wolves to attack. They, however, were intimidated by the elk's confidence, and none of them did anything.

I often saw dozens of elk grazing on grass near the Druid den forest. The elk seldom seemed concerned about the wolves, and it was rare for any Druid to chase them. Perhaps wolves regard elk that confidently hang out in their den area as animals that are too fast and strong for them to hunt successfully. As I had seen in the confrontation with the bull, wolves greatly respect confidence in a prey animal.

The bull bugled several times, and I realized his real agenda was fighting other bulls and romancing cows. To him, the pack was just a minor annoyance. He had more important things to do. The male yearling finally got a response when he got very close to the bull's rear end. The bull turned and lunged at the wolf with lowered antlers. That was the move that had fatally wounded wolf 6, but 107 easily dodged the thrust. At that point, the pack must have decided that the big bull was too much for them, because one by one they walked away.

I resumed watching the Druids and saw that they were continuing to hunt. They went into a nearby forest, and I got glimpses of them chasing elk, then saw wolves jumping up at one of the animals. The trees blocked my view, but soon it looked like the pack had killed the elk and were feeding on it. That was a classic chain of events for a wolf pack out on a hunt. They failed with one elk, kept on trying, and eventually got another one. That is what being a wolf is all about: determination to succeed, regardless of how many setbacks you go through.

WOLF 40 RESUMED her aggressive treatment of her sister. I saw her chase 42, pin her, and bite her on the rear end. After the alpha female walked off, 42 went to one of the black female yearlings and played with her. A few moments later, 40 ran over and pinned that yearling, eliciting yelps of pain from the young female. It looked like she was punishing her for being friends with 42. 42 went to her sister at that moment and licked her face, an act that distracted her from the yearling and ended her aggression. I wondered if that young female might remember the incident and later try to repay 42 for helping her.

40 later acted very aggressively toward another of the female yearlings. All three young females would be capable of having pups the following spring. Driving out some of them would give the alpha female's pups a better chance of survival. There was also the issue of how 40 was related to those yearlings. DNA testing later showed that 41, the sister she had harassed until she left the pack, was the mother of all three of them. Therefore they were 40's nieces. She might have viewed them more favorably had they been her

daughters. Years later, in another pack, the alpha female violently drove two nieces out of the group prior to the breeding season but allowed her two daughters to stay.

I thought about how 21 regarded 40's aggressive behavior to the other females and the disruption it created in his family. He had a laid-back personality, got along fine with the two male yearlings, and likely did not understand why the females were in conflict. When I saw him standing off to the side, witnessing 40 beating up one of the other females, I thought of a poster sold in many Montana and Wyoming stores at the time. It had a drawing of a weather-beaten ranch hand who looked annoyed. The caption under him read: "There were a helluva lot of things they didn't tell me when I hired on to this outfit!"

Soon after the alpha female's latest violent outburst, 103 and 106 chased several elk, and I was reminded of how dangerous elk hunting is for wolves. Closing in on one cow, 106 got right behind her. The cow kicked back and knocked her down. She rolled and tumbled but got back up. The wolf did not renew the chase, probably due to the pain from the blow. Over the years I saw many park wolves with bad limps or broken legs, injuries probably inflicted when elk kicked them. One alpha male broke his right front leg three times. It fused on its own after each break but ended up crooked. After each broken bone he soldiered on, walking and running on three legs until the bone fused.

Sometimes, as with the Crystal Creek alpha male, wolf 6, the encounters with elk were fatal. Late in October that year, we got a mortality signal from one of the Rose Creek male yearlings up the Lamar River. Several of us hiked out and found him. He had a single puncture wound in the center of his chest, the diameter of an elk antler point.

Hunting is not the only way wolves get injured or even killed. The Wolf Project has documented that the most common cause of death for adult wolves in the park is to be killed by a rival pack. Wolves fight over territories and kill each other during those battles, just like people fight and kill each other over land and possessions. As I wrote earlier, that aggressive territorial behavior works to limit wolf numbers in the park. Even though the park spans 2.2 million acres, an area nearly half the size of Massachusetts, there are only ten or eleven good-quality wolf territories with sufficient prey numbers to support a pack. Wolf packs in Yellowstone tend to be about ten animals. In recent years, the wolf population has averaged just one hundred, about the same number that originally lived in the area before it became a park.

In early November, I saw 42 and three of the yearlings on the north side of Lamar Valley. I also got signals from the Rose Creek pack from that same area. I soon spotted the Rose wolves, about five hundred yards north of the Druids. The two packs did not see each other. The Druids moved east, while the other wolves traveled west. I counted seventeen in the Rose Creek group, including the alpha pair. If the two packs had tangled, the four outnumbered Druids would have been in trouble, especially since 21 was not with them.

I continued to document how 21 was often ignored when he tried to choose the direction of travel for the pack. One day he wanted to travel east. The other wolves noticed 40 did not follow, so they stayed with her. 21 came back to them. Over a period of thirty-three minutes he tried to lead his family east eight times, then returned to them each time when they did not follow. When he went east the ninth time, the yearlings and the pup finally followed. Later 40 and 42 went that way as well. Before long, the alpha female passed 21, and she led farther east with him following.

I also kept track of when a Druid wolf initiated travel in a certain direction, and the rest of the pack followed. In the summer of 1998, 40 did that 48 percent of the time, while 21 did it only 20 percent of the time. Next in order was 42 with 17 percent and 106 with 4 percent. After the alpha female set the direction, other wolves might temporarily take over the lead position if she got distracted by something, but 40 had decided which way to go, and the others stayed on her designated route.

I spotted the Rose Creek wolves again on November 10 as I was getting ready to leave the park. This time I got a count of twenty-two. It was the largest pack in Yellowstone, almost three times bigger than the Druids. 9 was leading the pack and 8 was second in line. A big black wolf walked right behind 8 with a raised tail. It was uncollared and had a lot of gray shading on its coat, making it look a lot like 21, who had similar graying areas. That was wolf 18, 21's sister, the beta female, and last of the 1995 pups still in the pack. She had pups sired by 8 in 1997 and 1998, so some of the other wolves in the group were her offspring.

Wolf 8 had been with the pack for three years now, and he had proven himself to be a great alpha male and prolific breeder. Nineteen of the wolves in the pack, a combination of pups, yearlings, and young adults, were sons and daughters of 8. As far as I knew, he was the most successful male wolf in the park. His days of being the smallest wolf and getting picked on by his brothers were long gone. Those of us who had known him from the beginning were proud of him. He had achieved things way beyond what anyone could have expected.

When I reviewed my notes prior to leaving the park for the winter, I found that I had been in the field that summer and fall for 2,037 hours, the equivalent of fifty-one forty-hour weeks. Since I had been in the park for twenty-six weeks, I had averaged seventy-eight hours per week. I could well have been the hardest-working employee in the federal government that summer.

20

The Spring
of 1999

WHILE I WAS working in Big Bend in the winter of 1998/1999, I thought about how I was getting only part of the wolf story in Yellowstone by being there from May to November. To take it to the next level, I needed to study wolves in the winter, and I worked out an arrangement with Doug to stay in the park year-round. I planned to devote massive amounts of time, day after day, year after year, to watching wolf behavior and recording every detail of their lives. I was on a quest to understand the hearts and minds of wild wolves, on the individual level, then tell their compelling stories so other people could know them as well. I knew immersing myself in the wolves' world would involve many years of hard work under difficult circumstances but recalled what President John F. Kennedy said about setting goals: "We choose to go to the moon in this decade and do the other things, not because they are easy, but because they are hard." Now as I think back to that time, I realize I did not know just how hard that plan would be to carry out.

On April 30, 1999, my first full day in the park, I found five Rose Creek wolves on a new carcass north of the Yellowstone River. They moved off, and I lost sight of them heading toward the pack's traditional den site near Mom's Ridge. I later found out that wolf 18, 9's four-year-old daughter, was denning there. Her black coat was getting grayer with age, and she had a distinctive white spot on the top of her head. Her brother, 21, was developing the same white patch on the top of his head that she had, but on him, being male, it looked like a bald spot. The two wolves were nearly identical twins.

This was the third year she had had pups. She would likely take over as the pack's alpha female when her mother passed away. For now, 9 still held that position, and she had a new den site to the northeast. A younger female born into the pack in 1997 had denned west of Mom's Ridge. Eventually we learned that the alpha female had six pups, 18 had seven, and five were born to the new young mother. That was a lot of mouths to feed.

8, at five years, about forty-two in human years, was now just a year away from the average life span of a Yellowstone wolf. As I watched him on an elk hunt that spring and saw him trailing four younger wolves who were chasing the herd at top speed, I wondered if age was slowing him down or if he was just conserving energy and letting the others test the elk. When the young wolves caught up with a cow and attacked her, he ran to the site and helped pull down the cow and finish her off. 8 might be slowing down, but he still could pull his own weight.

Toward the end of May, sightings of the Rose wolves dropped off because they were hunting in higher country, so I concentrated on the Druid pack, which still had elk in areas near their den. The last day of that month, I went up Dead Puppy Hill and saw the Druids at their denning area near the Footbridge and Hitching Post lots. The alpha pair and 42 were in the group. 42 tucked her tail when she walked by the alpha female. I saw missing fur under the belly of 40, a sign she was nursing. A few days later, I also saw that her nipples were distended.

The sole surviving pup from the Druids' 1998 litter of two was now a yearling. He had been collared over the winter and given the number 163. We knew that he was a wolf that liked to investigate new things and heard he was getting in the habit of being around the road and cars. I watched a video of 163 sniffing a trash can in the Footbridge parking lot. He pulled trash out of the overflowing container, then casually walked around the lot and bedded down a few inches from the road. Later he picked up a piece of litter and swallowed it. We were worried that a tourist might toss food at him when he was near the road. If he ate it, he would probably start to approach people, hoping to get more food thrown at him.

The next morning, I found a bull elk that the Druids had killed during the night lying in the road. Wolf bite marks were on his throat. There were no other wounds, so an experienced older wolf had killed him, probably 21. A big male wolf measures about thirty-two inches at the shoulder, while shoulder height for a bull elk is around sixty inches. The elk's throat is a few inches higher. That meant 21 had to jump up twice his own shoulder height to grab the bull's throat.

I drove to the Lamar Ranger Station to report the carcass, then the other rangers and I dragged it off the road to a place where the wolves could safely feed on it. Around that time, I talked to a man who worked as a forensic wildlife investigator in Canada, and he gave me a better understanding of what happens when a wolf bites the throat of an elk. It will die of suffocation in one of two ways. The wolf could crush and rupture the jugular vein, causing blood to flow into the windpipe, and the animal would drown in its own blood. Or the wolf's powerful jaws could squeeze the throat so tightly no air could pass through it. Either way, death would be quick, within a few minutes. I would later find out that young wolves do not know how to make that type of fatal bite by instinct; they have to learn it by seeing older wolves demonstrate it. 21 had likely seen 8 make kills that way and was now imitating his technique.

Like his father, 21, 163 seemed driven to bring food to the pups at the den site. One morning I spotted him west of the den carrying a heavy elk spinal column with several ribs still attached. I lost him heading to the den forest. He was likely bringing it home to the pups so they could chew on the bones and play with them.

Two days later, I saw him trotting toward the den from a new carcass. I lost him in the trees and figured he was on his way to regurgitate meat to the pups. After twenty-five minutes he headed back to the carcass. On the way he ran into one of his older sisters. She licked his muzzle and he regurgitated meat for her. They ran off together and soon arrived at the carcass of a bison that had likely died of natural causes. After feeding, 163 made another trip to the den and pups. On that trip he paused to make two food caches. In the coming days, if the pack failed to make a new kill, he could go back to those caches, dig up the meat, and share it with the pups. That evening I saw 21 heading toward the den from the bison carcass with a very full belly. Fifty minutes later, he went back to the carcass for another load of meat for the pups.

The repeated trips the two males made from the carcass to the den and back, along with the food caches the yearling made, were good planning. The next day a large male grizzly took over the bison carcass. When the big bear left, a mother grizzly came right in with her three yearlings and further prevented the Druids from doing much feeding. At one point, 40 and 163 teamed up to drive the bear family from the carcass. As soon as the grizzly sow left, the two wolves ran in and gulped down meat as fast as they could.

The mother bear raced back but paused to check on her three yearlings. That gave the wolves more time to eat. The sow seemed conflicted between driving the wolves off and making sure her young were safe. Two more Druids arrived, and the four wolves harassed the four bears and managed to get to the carcass several times and do some feeding. After the bears had left, I saw 103, the smallest of the Druid adults, feeding on the carcass by herself. A big grizzly approached the site, but she refused to back off. Both ate a few feet apart without any problems. She looked tiny compared to the grizzly, but despite the size difference, she was not afraid of the bear.

It was not all business at the carcass, however. When 21 and his yearling son were feeding there later, they took time out to play. The big alpha male ran at 163, and the yearling, knowing this was a game, charged at him. 21 abruptly turned around and let himself be chased. Then he became the chaser. The younger wolf soon turned back, did a play bow, and ran forward, directly at 21, who was still running at him. At the last moment, he ran past his father and avoided a collision.

21 caught up with 163 and nipped him on the rear end. The two males playfully ran side by side, then sparred and wrestled. 21 must have held back on using his full strength for the match seemed even. Falling down, they continued to wrestle on a snowfield on the side of a hill. Both wolves slid downhill on the snow and rolled over several times. When they jumped up, the son chased his father again. As I watched 21, it looked like he was reverting to his carefree younger days when he played with his siblings or with 8.

ON MAY 15, I got an unexpected wolf signal to the north of Tower Junction. It was from former Druid male 104, who had left the pack the previous fall and joined the Crystal Creek pack as its alpha male. Why would he walk away from his new pack and come back to the north? The next day, I got a report that he was on the north side of Lamar Valley, heading toward the Druid den forest. I got his signal just west of the den. When I did a check twenty minutes later, the signal came directly from the den forest. I also got signals from three adult Druids there. As far as we knew, this would be his first reunion with his family since he left the valley eight months ago.

The following morning, I saw him at the Chalcedony Creek rendezvous site with one of his sisters. Both had spent a lot of time in that area when they were pups in 1997. I could identify him from a distance because he had a crook in his tail, just like his mother, 42. There was a tracking flight later that day, and he was located ten miles south of that area, about halfway back to the Crystal Creek pack's territory. Apparently, 104 was going back to them after visiting his relatives in Lamar.

AROUND THAT TIME, snowmelt caused the Lamar River to run at high levels. I saw 103 on the other side of the river, trying to find a safe place to cross. She waded across a small side channel, climbed up on a logjam that extended over the widest and deepest section of the river, and walked on the logs to the far side. It was a clever way to avoid a possible drowning.

A few days later, Jennifer Sands, a graduate student, and I hiked out to recent carcasses to collect information and samples: a tooth to age the elk and bone marrow to determine the overall condition of the animals. We left early when the river was at its low point for the day. By the time we got back to our crossing point, the water was too deep and swift to attempt wading. Remembering 103, we hiked to where she had crossed the small channel, waded through the water, then walked across the logjam to the other side of the river. If it had not been for her, we would have been stuck on the other side of the river for the night.

This little wolf was not only smart, but also the fastest wolf in the pack, the wolf version of sprinter Usain Bolt. I once saw her traveling with the alpha pair and 42. The wolves saw some elk and 21 charged at them. 103 joined in the pursuit and immediately passed the alpha male. Soon she was far ahead of the pack's adult females, as well.

Over the years I saw that female wolves, due to their lighter weight, were normally faster than big males. On typical hunts, a young female usually caught up with the targeted elk first. Her job was to bite into a hind leg and hold on, even if the elk kicked her in the head. Her weight would slow it down. If her sister ran in, she would grab the other hind leg. That would protect the first female from additional kicks. Then a large male like 21 would catch up, get out in front of the elk, turn around, leap up, and grab the throat. By themselves, the two swift females might not be able to kill the elk, and an average male, if he was on a solo hunt, likely could not outrun it. Wolves coordinate their attacks, each contributing what it can, to earn their dinner. However, I have seen exceptional females, regardless of their size, make fatal bites to the throat and kill elk by themselves. And I have seen faster-than-average male wolves. Just like human athletes, wolves exhibit a range of physical abilities.

The next time I saw the Druids leave the den, they spotted a herd of twenty-five elk near Chalcedony Creek. The wolves chased that group and several other herds for the next fifty minutes, but all the elk were too fast for them. After they gave up, one of the adults began mousing, hoping to get at least a small meal. Other Druids went back to an old bison carcass and scavenged. We had visited that site a few days earlier, so I knew there were only bones and fur there. I saw 163 gnaw on a big leg bone while two adults chewed on the skull. After getting what little they could from the site, the Druids tried chasing a new herd of elk, but still could not catch any. They went hungry that day.

As I observed the Druids, I kept records on how long the pack continued to return to old carcass sites to feed or chew on bones. I once saw them revisit a bison carcass a year after the bull had died and still manage to get edible bits off it. In late 1999, I saw them come back to a bull elk they had killed on July 8, 1998, over seventeen months earlier. Five of the nine wolves chewed on the skull, and one female picked up a leg bone, dug a big hole, and buried the bone in it, like a dog burying a bone in a backyard to gnaw on later. That would extend her use of the carcass past the seventeen-month period. I have often seen hungry wolves pluck fur from old elk hides and eat the leathery skin, the way starving people have been known to eat their leather shoes.

Wolves have evolved to go without food for substantial periods if hunting is poor. As an experiment in a captive facility many years ago that would not be considered humane today, an adult male wolf was not fed for nineteen days, and he survived. But wild wolves, if they have failed to make a kill for days, have the option of going out on another hunt, then another, until they succeed. The quality I most admire in wolves is their grit, a word defined as passion and perseverance. I think they love life so much that they cannot conceive of giving up.

WOLF PACK TERRITORIES in Yellowstone average about three hundred square miles. Bob Crabtree, who did coyote research in the park for years, told me a typical wolf territory might encompass ten coyote territories. As the Druid wolves traveled through the valley, they often encountered multiple coyote packs, and some of them would try to drive the wolves out of their family's territory.

One morning in the spring of 1999, I saw a male coyote chasing 42 near the Yellowstone Institute. She frequently turned around and confronted the coyote. Each time he backed off but continued to follow when she moved on. 42 kept her tail down to protect her rear end from bites. When she stopped to sniff a spot, the coyote ran in and nipped her tail. She turned around and chased him off, then moved on. The coyote followed and lunged at her tail again. 42 tucked her tail all the way under her belly, then turned back and drove him away.

I later witnessed the Druids get their revenge on that coyote pack. Five of them traveled to the area behind the Yellowstone Institute where those coyotes were denning. With 42 in the lead, they chased the coyote alpha female. I lost sight of the action, but I could hear the coyotes barking in alarm. People taking a class at the Institute saw 42 and then another adult female each dig at the den, reach down inside, pull out a lifeless pup, and walk off with it. I saw both of them chewing on the dead coyote pups.

Soon after that the wolves left the area. Bob's crew told me the coyote that had bitten 42 on the tail eleven days earlier was the alpha male of this pack. Since 42 had led the Druids from their den to that coyote territory, initiated the chase of the mother coyote, found her den, and pulled out a pup, I wondered if she had gone there deliberately to even the score. The coyote researchers later told me the adults moved two surviving pups to a new den site south of the road.

As I watched the Druid wolves and other packs, I saw coyotes repeatedly stealing meat from wolf kills. If the wolves left a carcass to bring food to pups, coyotes would swarm in and consume much of the meat before the wolves got back. In Chris Wilmers's Food for the Masses study, he saw up to sixteen coyotes at a time stealing meat at wolf-killed carcasses. Wolves probably regard coyotes the same way store owners think of shoplifters.

A FEW DAYS after the coyote incident, Doug Smith and other Wolf Project staff joined me at the Institute, and we hiked up a slope to the east. We were joined by Anne Whitbeck, a retiree from Colorado beloved by the wolf-watching community for her generous spirit and friendliness to strangers. We needed to investigate a site on that ridge. In early April, 42 had appeared to be denning up there. Her signals had come from a patch of trees near the crest of the hill. Those signals faded, got stronger, then faded again in a pattern consistent with a wolf going into her den and later coming out.

On April 9, Wolf Project volunteers Debbie and Jason spotted the alpha pair heading to that area, then saw 40 beat up her sister for four minutes, longer and more violently than usual. After that, both females and 21 went into the forest where 42 seemingly had her den and were out of sight for several hours. The signals from 42 indicated she was going in and out of her den. The next day the alpha female beat up her sister again. The bullied wolf later went to a fresh kill the other Druids had made but did not feed. Normally, a mother wolf nursing newborn pups would voraciously eat as much as possible at such a site. Something seemed to be wrong. A few days after that, she abandoned her den site, moved to the pack's main den, and helped 40 with her pups.

That sequence of events led people on the scene to suspect that 40 had gone into her sister's den and killed her pups, a disturbing thought. But for those of us who knew this wolf and her violent personality, it seemed like a definite possibility. She had driven her own mother out of the pack, as well as one of her sisters. Then she had gone after 42 and repeatedly beaten her up. We could well imagine 40 wanting all the pack's resources devoted to her litter, rather than sharing them with her sister's pups.

Our crew went up on that ridge and found 42's den in the forest. The tunnel went straight into the hillside and was wide enough for a person to crawl into. We found nothing there, no pup remains, but that was expected. The local coyote pack, the one the Druids had attacked, would have found and eaten any dead wolf pups at the site. In the end, we did not know for certain what had happened there, but 42's construction of the den, the pattern of signals indicating she was going in and out of it, and her abandonment of the site after 40 spent several hours there, were strong evidence that her sister did kill her pups.

How might that event have affected 21? The big male had bred both sisters in February, so if each had pups he would be the father. He had followed both females to 42's den site and had likely watched the alpha female go into the den. Up to that point, everything would have seemed normal to him. Female wolves other than the mother often go into a den to check on the pups. But 40 had probably killed her sister's pups when she crawled into the den. We do not know if 21 heard the killing or if he had walked off and was unaware of what was going on. Whether it was that day or a few days later when 42 abandoned her den, he must have eventually figured out that her pups were lost.

If the alpha female tried to do the same thing in the future, would he intervene and stop her? I had never seen 21 harm a female in his pack. He seemed to adhere to a code of behavior that prohibited him from doing anything that might injure a female, even if one bit him. During a later breeding season, I saw him get bitten repeatedly by a small female he was interested in. I tallied that she bit him nine times, snapped threateningly at him ninety times, and once even knocked him down. He snapped at her once and pinned her once. It appeared that he accepted her right to bite him and would not bite her back or use force on her. Later, after rejecting him, she ran off and bred with another male. That year, the Druid pack was very large, and the interrelatedness of the wolves was complicated. The female might well have rejected 21 because they were too closely related.

I tried to figure out how 21 and other male wolves came to have that code of conduct. While watching many den sites over decades, I saw clearly that the mother wolf was the undisputed boss of the family. If a pup wandered off too far, she would run over, grab it by the back, and carry it back to the den. Mother wolves are very decisive when it comes to correcting pup behavior, while the adult males rarely intervene.

Alpha females set the agenda for the pack: where to den, when to go out on hunts, and where to travel. As male pups grow up, they seem to retain that understanding of how wolf life works. The females make the decisions and enforce the rules. 40 was the queen; 21 just worked for her. If his code prohibited him from using force against her if she was about to harm another female's pups in the future, who could save them? The alpha female would attack any other Druid wolf that stood up to her. If a mother tried to defend her pups, I had no doubt 40 would kill her.

I later came across a passage in Jim Halfpenny's book *Yellowstone Wolves in the Wild* where he describes a similar event that took place in the spring of 1998. 40 was using the pack's main den near the Footbridge and Hitching Post lots, and 42 was based on the same ridge east of the Institute. As in 1999, wolf watchers saw the alpha female travel to her sister's den site, then heard fighting. Jim wrote, "From that day on, 42 never returned to her den." As in 1999, Wolf Project staff visited that site and found a den, but no remains of pups. Jim's information suggests that 40 may well have killed her sister's pups two years in a row. Perhaps what I had thought was a false pregnancy in 1998 had been the real thing.

21

Life at the Druid Den

B Y EARLY MAY 1999, many bison cows were calv-
ing, well before any elk calves were born. Two Druids
approached a cow and her new calf, reddish in color.
The cow drove them off. Two more Druids arrived and all
four surrounded the bison pair. The cow chased off the
nearest wolf. It came right back, and the wolves took turns
darting in at the calf, which was now crowded next to its
mother. The cow kept the wolves at bay by swinging her
huge head back and forth in a threatening manner when they
came too close. Then the cow and calf moved off. The wolves
ran after them, but the cow turned and faced them. The pack
never made a serious attempt to attack the calf due to the
effective protection of its mother.

Elk calves arrived in late May, and one morning I saw
the Druid alpha pair out hunting for them. The alpha
female charged at a cow elk, and I spotted her newborn calf

bedded down nearby. The wolves confronted the cow, but she charged at them, then chased the pair in circles. Picking out the biggest wolf, the cow ran at 21 and kicked forward at his hindquarters, a potentially crippling blow. He dodged the intended strike. As the cow concentrated on 21, 40 ran to the bedded calf, bit it, and struggled to pick it up. The mother raced over and drove the wolf off. As the elk dealt with her, 21 ran in, grabbed the calf, and ran off with it. Seeing that, the cow ran back and tried to kick 21. He had to drop the calf and run away to avoid her blows. But as she chased him, 40 ran in and got another bite in on the prone calf. The cow then got distracted by a nearby coyote. She veered off from 21 and went after it. That freed 21 to run back to the calf. He bit into it and that seemed to be the killing moment.

As pups grow, they need a lot of meat. It occurred to me that wolves may have evolved to time their breeding and birth of pups so they would be four to five weeks old, the period of weaning, when the elk and other local prey species were having their young in the spring. I documented that in one thirty-four hour period the Druid adults got at least four calves.

Being the most experienced, the alphas are usually the most successful hunters in a family, but on one occasion that spring, I saw the young Druid pack members succeed in a hunt when the alphas failed. An elk herd had spotted the Druids approaching and ran off. The pack chased them with 21 leading. He soon ended up in the middle of six cows and ran along among them. 40 and another wolf joined him. The trio stopped after failing to get any of the elk, then they looked back to the east. I swung my scope around and saw three of the younger pack members attacking a cow elk that was already on the ground. The alpha pair ran in and helped finish her off.

In late May, 21 left the den, crossed the road to the south, fed on a new elk carcass for a half hour, then moved back to the den to feed the pups. His protruding belly looked like it was bursting with meat. He was probably carrying a load of at least 20 pounds. It was Memorial Day weekend, the busiest weekend of the year for our section of the park. Soon he was just south of the den and heading toward the road crossing. Despite the No Stopping signs the rangers had put up, many drivers stopped anyway. Those cars were positioned directly between the wolf and his intended route to the den and the pups. Seeing the line of cars, 21 backed off and trotted west.

He paralleled the road but stayed in sight to park visitors. Soon there were a hundred cars stopped in the area or driving slowly back and forth. More vehicles arrived, and those drivers also stopped to watch and photograph the wolf. I could see 21 searching for a gap in the line of cars so he could run across to the north, but there were no breaks in what was nearly bumper-to-bumper traffic. There were too many cars already stopped in both lanes for me to use my red stop sign. I walked down the road, explained to drivers that a father wolf was trying to cross the road to bring food to his pups, and asked them to move on. But as soon as a few cars drove off, more vehicles arrived and took their place.

Four miles west of his intended crossing spot, 21 made a run to the road, but the line of cars stopped there caused him to turn back. He proceeded to dig a hole, regurgitated a big load of meat into it, and covered it up. He was stuffed with meat and likely feeling uncomfortable, especially due to the stress of not getting back to feed his pups, so he had to cache some of the meat. He went another two miles west and finally got across to the north. Now he had to walk six miles back to the den on that side of the road. That was the worst disturbance I had seen a wolf go through in Lamar. The incident showed how determined 21 was to get back to his pups. Nothing was going to stop him from bringing that meat home, certainly not having to walk an extra twelve miles.

IN EARLY JUNE, 104 was back, sniffing around an old bull elk carcass at the Chalcedony Creek rendezvous site and likely getting the scent of the Druid wolves there. Also in the area was 163, the pack's sole surviving yearling. We knew from DNA testing that they were cousins. 21 fathered 163, probably with 40, while 38 and 42 were the parents of 104. The former Druid, now two years old, had helped raise the yearling. What would the two wolves do if their paths crossed?

When 163 spotted his former pack mate, he moved away, looking over his shoulder frequently. 104 walked slowly after him. Neither wolf, it seemed, recognized the other. When the yearling broke into a run, the older wolf charged after him. As he closed in, 163 lowered his head and body into a submissive posture. At that point, he must have recognized his relative, because he wagged his tail. He licked the older wolf's face and the two began to play. After almost an hour and a half of traveling and playing together, the yearling moved off and swam the river in the direction of the den. I sensed he expected 104 to follow, but the older wolf was still south of the river when I had to leave because it was getting too dark to see.

The next day I found 104 alone at Slough Creek, sniffing at tall grass along the bank. A female duck, looking agitated, swam back and forth near him. I saw him reach down into some marsh grass and come up with a duck egg in his mouth. He ate it. He later found a carcass the Rose Creek wolves had been on, fed there the rest of the day, and in the evening moved off toward the Rose Creek alpha female's den. I got her signal in that direction, along with four other Rose Creek wolves.

I could not think of a good explanation for 104's behavior. He had left the Druids as a yearling and managed to join the Crystal Creek pack as their new alpha male, despite the bad blood between his family and that pack. Why did he leave those wolves when he was set for life as their breeding male? Now he was traveling alone toward another pack that considered his family an enemy. If the Rose wolves got his scent and tracked him down, they would attack and possibly kill him. Yet he confidently walked right toward them. As I lost sight of him to the north, I wondered if this was just another manifestation of his wandering spirit, always seeking out what was in the next valley or over the next ridge.

ON JUNE 11, I hiked up Dead Puppy Hill to watch the Druid den area. I saw 42 come out of the den forest, then noticed the pack's other six adults bedded there. Two-year-old male 107 had dispersed from the pack late the previous fall, and we had no idea where he was now. A gray pup got up and walked toward 42. That was my first sighting of a new Druid pup in 1999. More pups came into view, and I got a count of five: two blacks and three grays. The pups wandered off and the adults followed them at a slow pace.

The next day I saw six pups: two blacks and four grays. Two of the grays wrestled with each other, then all six tussled in a big pile. The pups were only about six weeks old and were already vigorously wrestling, looking like they had been doing that for some time. I eventually saw pups as young as three weeks wrestle each other. That is also about the time they are learning to walk, so wrestling is likely the first game pups play.

A few days later, two-year-old 106 was babysitting the pups. The pups had wandered away from the den area, and she was following them, carrying a long stick in her mouth. A pup ran over and tried to leap up and grab it but missed and fell over. 106 turned around and went back toward the den, still holding the stick, trying to lure the pups to a safer area. They ignored her, so she gave up and followed them once more, holding the stick over the heads of two pups. Both tried to jump up and snatch it.

The older sister tried going back to the den again, but none of the pups followed. At that point she sat down, with the stick still in her jaws, perhaps hoping the pups would come to her, but they ignored her and continued trotting east. She dropped the branch and followed. Yearling 163 joined them. I had seen him earlier with a dried-out bison dropping, shaped like a Frisbee, in his mouth. A pup near him was now walking around with the unlikely toy.

The pups tired of their exploration and headed back to the den, led by one of their own rather than an adult. 163

seized the moment, got out in front of them, and led them toward the den forest, with the bison dropping now back in his mouth. When the pups stopped following him, he ran back and dropped the scat in front of them. A gray pup picked it up, then 106 ran in and romped with the pups. When 163 saw the bison dropping on the ground again, he picked it up, shook it like it was alive, and walked off with it. A gray pup followed him, grabbing the scat when 163 let it fall to the ground.

163 gave up on leading the pups back to the den and played with them instead. He found the stick the female had used to try to get the pups to follow her. He ran around with it and passed a black pup. It chased him. The yearling turned around and did a play bow to that pup. He dropped the stick and watched as the pup ran toward it. Just as the pup was about to reach it, 163 grabbed the stick and ran off. Soon he dropped it again and ran circles around the pups. He looked like he was having the time of his life.

It wasn't all play, however. 163 was also monitoring the pups. A bit later I saw a gray pup trip and fall. Two other pups ran over, pounced on it, and seemed to be treating the gray very roughly. 163 ran there and hovered over the pups, acting like he was trying to stop the harassment of the first pup. But that pup jumped up and gleefully romped off like it enjoyed the rough play.

As 163 continued to play with the pups, chasing one, then another, a pattern emerged. Just as he was about to catch up with a running pup, it would abruptly collapse and act submissive. The sequence happened five times. That instinctive behavior might save a pup if a wolf from a rival pack chased and caught it. Years later, I saw a pup do exactly that when an alpha female from a neighboring pack pursued it. The pup collapsed and went limp while she nipped it a few times. Then she stopped and walked away without doing any real harm. It looked like the pup's submissive behavior short-circuited her aggression. Even though all this activity looked like play, lessons in wolf social behavior were being learned.

IN LATE JUNE 1999, from my vantage point on Dead Puppy Hill, I watched the six Druid pups go down to the marsh for the first time. They were being supervised by 40 and 106. After a lot of play and some exploration, all the pups followed the alpha female in single file as she moved back uphill toward the den forest. It was like watching a line of nursery-school children being led back to the classroom by their teacher after recess. And, just as with small children, some pups moved along faster than others.

For both pups and children, there can be dangers in being left behind. One day the six pups were running after 42 toward the den forest, and a small black pup was having a hard time keeping up. A black bear with two cubs appeared after 42 and five pups had gone into the trees. 42 came back out, saw the bears, and went to that last pup. They walked side by side uphill toward the den forest, away from the bear family. 42 licked the pup as they traveled. When it got distracted and ran back toward the nearby road, 42 raced downhill, blocked the pup from going that way, then led it uphill again.

Usually, the young adults were enthusiastic participants in pup activities, but not always. A few weeks later, four pups had a play session. When 103 walked in, they pestered her for a feeding. Perhaps because her stomach was empty, she ran from them, but the pups chased her and continued to demand food. Soon it looked like the pups had invented a new game: wolf pinball. She would flee from one pup but run into another one who also pestered her. When 103 ran from that second pup, she was blocked by a third. She bounced from one pup to another.

163 continued to spend a lot of time playing with the pups. I would see pups chasing him as he ran from them with a stick in his mouth. He would pick up an antler and show it to the pups, and they would chase him. He was so much larger than they were that when he wrestled with the little pups, they looked like the small stuffed animals dogs like to carry around. And yet he was gentle with them and held back on his strength, so he appeared to be evenly matched with them.

We were getting more concerned with 163's casual attitude toward human-related things. He often walked down the park road, passing by parked cars within a few yards. If he saw litter on the side of the road, he sometimes picked it up and carried it around. Those traits classified him as a habituated animal, meaning he had gotten so used to roads, cars, and people that he regarded it as safe to be close to them.

One day I found him walking along the road within six feet of a stopped van with excited and noisy people inside. I used my park radio to call in a law-enforcement ranger, and Mike Ross soon arrived on the scene. With his flashing lights on, Mike got out and yelled at the wolf. Now a bit frightened, 163 ran off the road to the north. To make the lesson clearer, Mike ran after him, still yelling and waving his arms.

The Park Service calls that aversive conditioning, and it can change an animal's behavior the first time it is used. But some individuals get used to that and ignore the yelling and chasing. In those cases, a ranger might hit the animal on the rear end with a rubber bullet, hoping a painful experience near cars and people will teach a wolf to avoid them. I monitored 163 after that, and Mike's treatment seemed to be working. A few days later, he was south of the pack's den. He saw four people on a nearby hiking trail and ran away from them, looking over his shoulder as he fled.

Later that day, 163 went back toward the den and waded across Soda Butte Creek. Pausing midstream, he looked down into the water, reached into it, and came up with an old rubber boot. After carrying the boot in his mouth for a while, he dropped it back into the water. As he moved toward the road, I saw ten cars were stopped between him and the den. Twenty-six people were out of their cars photographing him. Pausing to look at the crowd, the wolf tried to circle around them, a good sign of his changing behavior toward humans. He crossed the road, and I lost him heading up to the den. A few days later, he went back to that section of the creek, waded out, found that boot again, and carried it up to the den to give to the pups as a toy.

In mid-June, I got signals from 104 from the Chalcedony Creek area but did not spot him. I could not figure out what he was doing, but after so much time away from the Crystal Creek pack, it seemed he was not going back to them. In late June, I saw him with 163 near a carcass up the Lamar River. Other people told me that the two wolves had fed side by side at the site. Later in the day, signals from those two wolves and from 21 came from that area, suggesting 21 had joined them.

21 continued his play sessions with 163. One day I spotted him pretending to be afraid of his son. He ran from him with a tucked tail, acting like 163 had just defeated him in battle, had taken over the alpha male position, and was driving him out of the territory. The yearling caught up with his father, wrestled with him, and pinned him. Jumping up, 21 ran off, but the yearling soon caught him and once more pinned him. It was extraordinary to see the huge, powerful wolf deliberately lose to the smaller male.

Why did 21 seem to take great joy in playing with his son like that? I wrote earlier about how oxytocin hormones are released when fathers and sons engage in roughhouse play and how that enhances emotional bonding between them. I think that was what happened when 8 and 21 played that way, and I had just seen 21 repeat that experience with his son. I felt I had witnessed a profoundly intimate moment between two wild wolves.

I know a little bit about that subject. My father, Frank, was an engineer and World War II veteran, a quiet man who never spoke about his feelings or emotions. I have no memory of him ever saying he loved me or hugging me. Since he died when I was ten, I have very few recollections of him. There is, however, one memory that has stayed with me. One afternoon, both of us were alone in the living room. My father got up, walked over to me, and did something I never would have imagined he might do. He asked if I wanted to wrestle with him. Stunned, I said yes. He got down on his hands and knees and we wrestled.

Although I was only about six at the time, I knew he had recently had a major heart attack, and the doctors had told him not to engage in any strenuous physical activity for fear it might set off a fatal second attack. Despite that, he wanted to pretend we were in a wrestling match. After going back and forth a bit, he maneuvered me into a position on top of him, then declared that I had pinned him and won the match. He got up, went back to his chair, and resumed reading the paper.

When I think of my father now, I understand that he grew up in an era when dads usually did not have close emotional relationships with their sons, but he made the effort to wrestle with me that day and let me win, his way of showing he cared about me. He could not put it into words, but he could do that. I will always have that vivid memory of him, and it is enough for me. And that experience is what makes me identify so strongly with the roughhousing play wolf 21 engaged in with his son.

THE PUPS STARTED mousing in the marsh on July 14. A few days later, I saw one of the gray pups walking around with a vole in its mouth. The pup bedded down and began eating it. Two other pups were nearby, also hunting for voles. 21 came into the area from a trip to the south and saw them. Totally concentrating on vole hunting, they did not pester him for a feeding or even look up when he walked by. 21 went to the gray pup with the vole and sniffed it. The pup glanced up at him with part of the vole hanging out of its mouth, then resumed eating. 21 walked off, and I wondered if he was pleased to see those pups taking the initiative to learn how to hunt on their own. It was the beginning of a lengthy process that would culminate with those young wolves being assets to the pack when hunting elk and other prey.

I soon saw all six pups mousing in the marsh, concentrating on listening for voles and scanning the grass for them scurrying around. Later 21 returned to the marsh and, perhaps inspired by the pups, soon had a vole hanging out of his jaws. 163 came over to greet him, and 21 dropped the rodent in front of his son, who grabbed it. Then 21 walked over to the pups and watched as they moused. After that he bedded down farther uphill and monitored the pups, looking like a human father watching his children play.

Like any father, 21 had to put up with a lot from his young sons and daughters. When the pups pestered him, they learned to ignore his growls and snaps, knowing the threats were never carried out. I saw three pups chasing him one evening. He turned around and snapped at the lead pup, missing it by a few inches. The next pup ran up and licked his father on the face, unconcerned about the threatening lunge it had just witnessed. 21 snapped at that second pup but deliberately missed it, just as he had with the first one. The three then aggressively begged for food. 21 sat down and endured their harassment. When the pups finally got bored, they ran off, leaving him be.

Early one morning the alpha pair, 42, and 106 left the den forest, ran down to the road, and crossed to the south. A black pup followed them but lagged behind. The adults continued south at a fast pace, crossed the creek, and moved on, unaware of the pup. It trailed their route over the road, continued to the spot where they had crossed the stream, and

waded to the other side. That was the first time that year I
had seen a pup south of the road or creek.

The four adults continued southwest, toward a new carcass, and disappeared from sight. The pup trotted in that
direction but veered off their route. When it got to the high
bank above the Lamar River, it sniffed around, searching for
their scent. Then it looked around to see if it could see them.
The young wolf seemed perfectly calm, despite being only
three months old, roughly equivalent in age to a three-year-
old child. At that point the pup headed back north, following
the exact route it had just taken to the south. It must have
been following its own scent trail.

With surprising confidence, the pup trotted through the
sage, waded the creek, crossed the road, and walked back
uphill to the den. The little Druid had a good sense of direction and needed no help in finding its way home. I looked
up the slope and saw 105 standing at the top, looking downhill. She must have spotted the pup and was monitoring its
return from its big adventure. She did not run down to the
pup and lead it back to the den, but let it figure out the way
home by itself. The pup's round trip had taken two hours.

As the pups became more skillful in hunting voles, I saw
them practice catch and release. One of the grays caught a
vole, let it go, then grabbed the rodent again as it tried to
scurry off. Later I saw a pup catch a vole, shake it, toss it
in the air, then pick it up. After shaking it and throwing it
up again, another pup raced over, intending to steal it. But
the first pup saw its sibling coming, snatched the vole up,
and ran off. When the pup stopped and dropped the vole, I
saw the rodent was still alive and trying to get away. As it

disappeared into the grass, the pup frantically chased it and got it once more.

The pups learned how to tackle each other. I saw a black pup chase a gray, catch up with it, reach forward, and grab one of its hind legs, forcing the other pup to stop. Later, when the black chased another gray, it bit the gray on the rear end, pulled it down, and stood over it in the dominant position. Then the black gave the pup another bite on its rear end. I already had a list of games wolf pups and yearlings played. It included ambush, catch me if you can, catch and release, snow sliding, sparring, tossing and catching, tug of war, wolf pinball, and wrestling. I added tackling to the list.

By late July, the pups were roaming far and wide in the den area. As I watched from Dead Puppy Hill, I saw a black and two grays moving south toward the road. One of the grays looked at the people and cars in the Footbridge lot and ran back north. The other two pups did the same. Their response indicated the young pups had an instinct to be afraid of people. That is a good thing, for the border of the park was only ten miles north of the Druid den and fourteen miles to the east. In later years, when wolves were taken off the regional endangered/threatened species list, legal wolf hunting would be allowed in the three states surrounding the park, so any Yellowstone wolves that viewed humans as nonthreatening might be easily shot by a hunter. 163 would have been like that if Mike Ross's aversive conditioning had not worked.

22

Moving On to the Rendezvous Site

EARLY ON JULY 27, I spotted 21, 105, and 163 south of the den, across the road. They were looking back to the north. A few minutes later, all six pups ran to the three adults and greeted them. They must have followed the adults across the road. The pups led south and seemed eager to explore new country. The nine stopped and howled. We heard answering howls from other pack members farther south. After looking that way, 21 ran south and the others followed. The pups were right behind their father, while 105 stayed last in line, probably to make sure no pup strayed off their route. 163 could not resist a quick game. He spun around, lay down, then jumped up and ambushed a black pup when it caught up. The pup squirmed out of his grasp and continued to run after 21.

All nine wolves trotted in single file along a hiking trail the adults often used when traveling through that area. After losing sight of the pack, I drove a mile west and walked up a hill that gave me a better view. The whole pack, seven adults and six pups, were now west of the Lamar River heading toward the Chalcedony Creek rendezvous site. That meant the pups had not only crossed the road, but also successfully swum the river.

I later saw that pups naturally know how to swim, but some are fearful when they first encounter a river or wide creek. 42 was a master at taking control of that type of situation. I once saw her swim back to pups who refused to enter the water, pick up a stick, show it to those pups, and run off. As they chased her, she waded into the creek, still holding the stick, and the pups followed, not understanding she was tricking them. When it got deeper, they started to dog paddle and were halfway across before they realized they were swimming. 42 seemed to be gifted at figuring out how to help pups overcome problems in ways that gave them confidence. She could have picked up the pups one by one in her mouth and carried them across, but that plan would have never taught them the swimming lesson.

The next test of the pups' mettle was a herd of thirty-six bison. 21 and 106 were calmly leading the pack past the herd, but the pups hesitated when they saw the enormous animals. 21 stopped and 42 ran back to the pups. I could see only one of them now, and it was running back toward the river. 21, sensing something was wrong, also went back. Soon all of the adults ended up on the high bank above the river, looking down toward the water. They then did a group howl. That was an effective tactic, for the spooked pups ran to the sound. After a greeting, 21 led west. The pups followed immediately. This time he chose a route that took the pack well away from the bison herd. One pup paused briefly when it saw the herd, then continued on. 21 also paused and watched the pups march forward, giving them time to be comfortable about passing the bison.

No longer interested in the bison, the pups explored the Chalcedony Creek rendezvous site. 42 joined them as they sniffed around the area. 21 monitored them, then led the pack farther west. The six pups followed right behind him with the other six adults trailing behind, which meant they could round up pups who wandered off the travel route. One of the gray pups passed 21. The little male seemed totally comfortable exploring new country and leading the group. He looked like a natural-born alpha. Later, 21 took over the lead once again, regularly looking back to check on the pups.

When some of the other adults lay down at the rendezvous site, the pups joined them, and 21, realizing they were tired of traveling, turned around, and bedded down among them. Before long, big brother 163 was playing with the pups. He grabbed what looked like a long branch and showed it to the pups, and they chased him. Several pups caught up with him and grabbed the stick. They all tugged one way while their older brother pulled in the opposite direction. As they played, I saw what they had was not a stick but cargo netting. It must have been left behind during the park's 1988 fires when helicopters carried supplies to ground crews in that type of netting.

The pups later explored the surrounding terrain and found a marsh where they could hunt voles. 21 seemed especially attentive to the pups. I saw him go to a black pup and greet it. The pup ran off in a playful manner and 21 followed it. The pup repeatedly looked back, making sure 21 was still following, then ran on again. It was an endearing sight to see the huge alpha male protectively trailing the little pup around the area. He was now four years and three months old, about the same age as a man at thirty-seven.

The Chalcedony Creek rendezvous site was a great playground for the pups and a relatively safe area for them when the adults were off on a hunt. After making a kill, they had a much easier time bringing food back to the pups since most of their hunting routes from this site did not involve road crossings. Once again, 21 made the most trips back to the pups from a new carcass and carried more food to them than any other pack member. After having witnessed the care and attention 21 lavished on the pups at his mother's den in 1997, I was sure then that he would be an ideal father when he had pups of his own, and now I could see that he had become exactly that.

IN EARLY AUGUST, a grizzly sow with three new cubs came into the rendezvous site. The mother bear moved toward the bedded wolves without realizing they were there and her cubs followed. 40 and 105 got up and ran toward the bears, then the alpha female stopped, bedded down, and watched them. The sow stood up and looked at her, then gathered up her cubs and moved away. 40 followed. 21 went to check on the pups, then walked toward the alpha female and the grizzlies. As he continued toward the bears, 21 alternated between looking at them and turning around to monitor the pups. 40 was now stalking the bears.

The sow and her cubs ran toward the river. The wolves had a kill in the river corridor, and the mother bear had probably gotten the scent of the carcass. All the bears went out of sight in the direction of the kill site. I checked on 21 and saw him heading back to the pups. He bedded down with them, ready to protect them if the grizzlies returned. His actions made me think he was more concerned with guarding the pups than defending his elk carcass. I saw that 40 was also going back to the pups. 21 waited for her to arrive and bed down among the pups before he got up and headed toward the bears. Now that she was watching the pups, he could deal with the grizzlies.

I lost 21 going to the carcass site, then heard from another observer that the grizzlies had just run out of that area. Right after that, I saw 21 near the kill. He must have charged the bear family and driven them away. Later he walked back to the pups and regurgitated meat to them. When the other adults returned from visiting the carcass later that day, the pups were too stuffed to eat another helping. As the day went on, 21 and other adults periodically walked around and ate pieces of leftover meat. One female buried some of that unused food to hide it from ravens so she could eat it later when she got hungry.

I later spotted a male pronghorn coming into the rendez-vous site. A black pup saw the pronghorn, jumped up, and charged at him. The pronghorn sped off at perhaps sixty miles an hour. Three other pups saw the black chasing him and joined in. They soon gave up, realizing they were hope-lessly outmatched.

After that, I saw a touching moment between the alpha female and her sister. 42 went to the bedded 40, lay down beside her, and licked her face. She then lifted her front right paw and gently put it up by her sister's face. I saw 40 lick that paw. When she stopped, 42 put the paw up by her face again and she resumed licking it. I later noticed that 42 was limping on that leg and wondered if soliciting the licking from her sister might help heal her wounded foot. I had often heard that dogs have an antiseptic chemical in their saliva, and I found a research report that proved their salvia kills two common forms of bacteria that infect wounds: *Escherichia coli* (commonly referred to as *E. coli*) and *Streptococcus canis*. If dogs have an antibacterial chemical, wolves would have it as well.

ONE MORNING THE pups all stood up and looked west. I swung my scope in that direction and saw a mother grizzly with three yearling cubs. After a few moments, the pups bedded down and seemed unconcerned about the bear family. By that time, to the pups, grizzlies were just neighbors. They also paid little attention to the big bison bulls that frequently walked through the area. When one huge bull came their way, nearby pups played together and ignored him. At other times, pups followed big bulls around the meadow. I saw two trail just ten yards behind a bull. When he flicked his tail to drive off flies, the startled pups ran. But they soon came right back and got even closer to him. He flipped his tail once more and they ran off again.

On another day, I spotted 163 returning to the rendezvous site with a strip of meat. Four of the pups ran after him. One tried to snatch it out of his mouth, but he blocked it with his shoulder. He laid the five-inch piece of meat on the ground in front of them, like he was daring them to pick it up. One pup crouched down and tried to reach it, but the yearling lunged at it and the pup backed off. 163 picked up the meat, walked off a short distance, then put it down again. The pups came to him, but did not try to get it. After grabbing it and walking away a second time, the yearling once more laid it on the ground. The pups joined him. Two made attempts to steal it but he blocked them. All four pups bedded and watched him, looking like schoolkids waiting for their teacher to give them permission to eat a snack. Their situation reminded me of how some dog owners put a treat on their pet's nose and train the dog to stay motionless until they give it permission to eat it. Dogs must hate that trick.

I had seen 163 play a similar teasing game with the pups back at the den when they were younger. I thought maybe both games would teach the pups about the principle of ownership. After four minutes, 163 turned around and walked off without the strip of meat. The pups had obediently stayed bedded down a few feet from the meat for that duration. As he trotted away, a gray ran in and grabbed it. 163 paused to look back at that pup, then continued on. The other pups came over, but none tried to steal the meat from the first pup. They respected its right of ownership.

That gray pup fed for a while, picked up the meat, walked off, then looked around for a spot to bury it. He dug a shallow hole, dropped the meat in, and pushed dirt over the hole with his nose to complete the process, the same sequence I had seen adults go through when they cached meat. That was the first time I had seen a pup so young make a cache.

The use of the nose to cover the cache intrigued me. Why would a wolf willingly get its nose dirty by pushing dirt over a hole rather than using a paw? I later examined a fresh wolf cache and saw the site was hard to spot. It looked no different from the surrounding area except for a small amount of fresh soil. But nearby squirrel burrows also had fresh dirt. How could a wolf, especially a pup, later find that cache? Perhaps when it pushes dirt over the hole with its nose, a wolf memorizes the scent of that exact spot and can easily find it again later.

I noticed 21 was consistently more playful with 42 than he was with the alpha female. I saw him go to 42 with his head lowered, as though he was pretending to be subordinate to her. He crouched down in front of her and rolled onto his back. She stood erect over him in a dominant position, her raised tail wagging. 21 lifted his head and licked her face, something a pup or yearling would do to a superior animal. Then they romped off together. Catching up with him, she jumped on his back, and he collapsed under her in a submissive position. It looked like 21 was letting 42 pretend to be an alpha female that day.

IN MID-AUGUST, WHEN the Druids were out at the rendezvous site, I got signals from 104 to the west. I spotted him a few minutes later, trotting east toward the Druids. The wind was blowing from the west, so the pack would soon get his scent. A few minutes later 21 and 42 moved in his direction. The alpha female joined them and took over the lead position. 163, one of the younger females, and three pups followed. I saw 104 going toward them, unaware of their presence due to the direction of the wind. 40 was heading right to him, but thick sage blocked her view. I had seen

104 with 163 on a carcass at the rendezvous site in late June and had watched the two brothers playing with each other there earlier. In mid-May, I had spotted 104 with 105 and got his signals from the Druid den forest. All that indicated he was on good terms with his family back then. But this was mid-August, six weeks after his last known contact with them. Would they still be friendly to him?

104 veered southeast and passed by the Druids without seeing them or the others spotting him. The Druids still were looking west, where he had been. Now that he was east of them, the wind took their scent toward him. He must have gotten that scent for he stopped and looked in their direction. I scanned west and saw the Druid alpha pair out in the open. 104 should have seen them too and figured out this was his family. But he turned around and ran east, like he was fleeing from an unknown pack. He stopped, looked back, and ran east again four more times. On his fifth look back, he realized the other wolves were not pursuing him. As he continued to trot east, we lost him in the river corridor. I later saw him going up Cache Creek, on the east side of the Lamar River, far away from the other wolves.

The Druids returned to the rendezvous site with 21 leading. They had not seemed concerned about getting 104's scent and probably would not have had a problem reuniting with him. I later wondered if the scent of the new pups might have confused 104. He could have recognized the scent of the other wolves, but possibly not the smell of the six pups. Did those scents make him wary, thinking unknown wolves were in the area? If that was the case, he played it safe and left.

THE MOTHER GRIZZLY with the three new cubs continued to visit the rendezvous site. One evening, the bears
ended up south of where the wolves were bedded down in
tall sage. The sow must have gotten their scent for she stood
up, looked their way, then ran west. Her cubs followed, also
at a run. 40 got up, saw the bear family, and ran toward them.
Soon she crouched down in thick grass and watched them
from that hidden position. She was about 150 yards away
from them. When the bears went out of sight behind a hill,
40 stalked them, keeping the hill between her and the grizzlies. They came into view again, just thirty-five yards from
her. Since the wolf was in dense sage, the bears could not
see her.

As the grizzly family went west, 40 followed in stalking
mode, slowly putting one paw down, then another, as she
stared at them with her head lowered. The last cub in line
stood up and looked back her way. It was twenty-five yards
from her. The cub must have seen the wolf for it ran to the
mother bear. The sow realized something had scared her
cub, but she could not see the wolf, who was now crouching
motionless in sage the same color as her coat.

The bears continued west and 40 followed. When they
ran, she ran, and when they slowed down, she did as well.
Each time the sow turned to look back, the wolf froze. Soon
she was within twenty yards of the last cub in line. A few
moments later she ran forward and bit that cub on the rear
end. She immediately turned and ran back to the east. It
seemed like it was a game to her to see if she could count
coup on one of the bears. All three cubs ran to the sow, who
stood up and looked around for whatever had scared them.

The wolf was hiding once again in the sage, just fifty yards away, and the mother could not see her.

In another stalk, 40 slowly crept toward the family and got within ten yards of a cub. At that point, she had to cross an open area in the sage. The cubs turned around at that moment and looked her way. The wolf froze. Even though she was in plain sight, they did not see her because her coat blended into the background so well. When the grizzlies moved on, the wolf resumed her stalk.

Choosing her moment, 40 charged at the cubs through an open area. They saw her and ran toward the sow. Turning around, the mother bear spotted the wolf charging at her cubs and ran back to them. As the wolf came at them, the cubs pressed against their mother. 40 paced back and forth in front of them, just a few yards away. The sow seemed calm and confident in her ability to deal with the wolf. She stood by her three cubs. If she had charged at the wolf, her cubs would have run after her. Then 40 could circle around, get behind her, and bite one of her cubs. The sow was doing the right thing by staying in place with her cubs by her side.

The standoff continued for some time, then fog rolled in, and I lost sight of the animals. When it cleared later, the bears were marching off in single file. I did not see the wolf. She must have been satisfied with her adventure and returned to the pack.

23

The Stubborn Pup

ON AUGUST 12, the seven adults and six pups in the Druid pack were bedded down at the Chalcedony Creek rendezvous site. 40 got up and started to move east. With the exception of 105 and 106, all the wolves followed her. At times, a gray pup overtook the alpha female and led the pack. As the wolves approached the river, the bigger of the two black pups, a male, turned around and went back to the rendezvous site. He rejoined his two older sisters who had stayed there. The other five pups continued to travel with the adults.

I got glimpses of 21 leading the pack south, toward the Specimen Ridge Trail, a route that takes hikers from the floor of the valley to the top of that high ridge. The adult wolves often hunted up there, an area where elk herds spend the summer months. I lost the wolves as they entered the trees.

The next morning Doug did a flight and found the main Druid group high up on Specimen Ridge.

The two females left the rendezvous site early the following morning, August 14, and went up the ridge to join the others. The black pup was alone at the site but seemed fine. He was hunting voles. Later that morning, 42 came back to the rendezvous site and the pup ran to her. 42 must have realized one of the six pups was missing and came back to look for it. She regurgitated meat to him. When the hungry pup pestered her for more food, 42 regurgitated a second time.

She set out along the route the two female adults had taken earlier in the day and the pup followed, but then saw or heard something off to the south and ran in that direction instead. The pup pounced for a vole, missed it, and continued south, hunting for more rodents. He seemed to have lost interest in following 42. Voles were more important to him in that moment. 42 stayed at the rendezvous site all day and monitored him. The two wolves did a lot of howling, trying to contact the other Druids, but they got no responses.

When I scanned the area early the next morning I saw three females with the black pup: 42 and two of his older sisters, 105 and 106. He howled frequently, but the adults did not join in. 105 regurgitated meat to the pup. The adults were keeping him well fed. I also saw him catch and eat voles that day. Over the next few days, various adults came to the rendezvous site to check on the lone pup and feed him. He never followed any of them when they left. He seemed to like being there and wanted to stay, even without the company of his five siblings.

One day the same two young females were tending him, but neither had any food for him. I saw 106 dig at a nearby ground squirrel colony, then dash after a squirrel when it fled from another entrance. The wolf pounced and got it. She ate most of it and walked off. Her sister came over and fed on what was left of the squirrel, then went to the pup and regurgitated the meat to him.

The pup was having a lot of success hunting voles. One evening he got three voles, then went to the bull elk carcass left over from June and chewed on bones. He was capable of feeding himself to some degree.

Five days after the other pups had followed the adults up to the top of Specimen Ridge, I found the alpha pair and 163 heading back toward the rendezvous site without the other five pups, who were probably still up in the high country. The black pup ran to the alphas and they greeted him. 21 walked off. The pup did not follow. 40 joined 21 and both went on the route toward the river. They soon stopped and looked back, expecting the pup to follow, but he stayed back. 163 went to the pup, his tail wagging, and the two had a friendly greeting. Then 163 walked to the alphas, who had bedded down, and he and 21 romped around playfully for a few minutes. Most pups would have run over and joined in the play, but the black pup still refused to follow.

Soon 21 led the adult wolves farther away from the rendezvous site. The pup did not follow. His three family members disappeared into the forested lower section of Chalcedony Creek. From there, they could take a game trail to the top of Specimen Ridge, where the rest of the pack would be. I saw the pup sniff where the alpha pair had bedded down, then look over to where they had gone out of sight. Then he began mousing and concentrated on that for over an hour. I saw him catch and eat several voles and grasshoppers. He then found an old elk rib bone to chew on.

In the evening, 42 arrived at the site and the pup ran to her. When she later followed the scent trail of 21 toward the forest, the pup focused on hunting for voles and insects. Instead of heading up to the rest of the pack, 42 came back and stayed with him. Both wolves were still at the rendez-vous site the next morning. 42 started to leave and tried to get the pup to follow her, but he refused. As she got farther away, the two howled at each other. He still would not fol-low. Soon I lost her signal, indicating she was out of the area and likely heading up the ridge. The pup was alone again, by choice. I saw him that evening and during the next two days, still by himself.

On August 20, the tracking flight found the Druids at an upper section of Opal Creek, a high-elevation rendezvous site that was much closer to the elk herds in their summer range than the site at Chalcedony Creek. The next morn-ing the pup howled as he looked that way, and I heard faint answering howls. He ran in that direction, proof he had heard them as well. After pausing to howl again, he ran far-ther that way. Stopping once more, the pup howled. I could hear him clearly from two miles away. The other Druids up at the new rendezvous site would be about three miles from the pup, close enough for him to hear their howls and get the correct direction. He ran off that way and I soon lost him. I felt good about the pup now. He finally had enough of life

by himself in the valley and was going to his family up in the
mountains. He had been alone for seventy-two hours.

Doug flew on the twenty-sixth and later told me he had
seen the Druids at the Opal rendezvous site. There were four
adults and five of the six pups, including both blacks. That
meant the black male had traveled up the mountain range
by himself and reunited with his family, despite never having been there before. He must have followed the scent trail
of other wolves, an impressive accomplishment for such a
young pup. Doug saw three gray pups that day. There should
have been four, so one was missing.

During the flight, Doug also found 104 at the south end
of Lake Yellowstone, less than a mile from the Soda Butte
pack. To get there, he would have passed through the territory of the Crystal pack, the group he had joined after leaving
the Druids. He either passed by them or met up with the
wolves, then continued farther south. We still could not figure out what he was doing.

The September 1 tracking flight saw 42 and a black pup
at the Opal rendezvous site. That same day, 105 went to the
Chalcedony site and seemed to search for pups. The next
day, 42 and another of the two-year-old females came back
to that area and also looked for pups. Both females howled.
105 returned there on September 4, howled, and scanned for
missing pups. The adult females were clearly worried. I was,
too. Any wolf pup away from adults would be easy prey for a
bear, a mountain lion, or a coyote pack.

There were no Druid signals in the valley on September 5.
Bill Wengeler, a seasonal Park Service naturalist, and I hiked
up the Specimen Ridge Trail to look for a viewpoint where

we could see the Opal Creek area. We left the Footbridge parking lot, followed the trail to the Lamar River, waded it, and continued uphill for several miles. It took four hours to get high enough to see the Opal Creek area. I got signals from the pack that way. To get a better view, we walked another mile, then stopped on a knoll and set up our scopes.

I found what looked like the rendezvous site, about two miles away. It was a meadow surrounded by a thick forest. The only prominent feature was a low hill with a single conifer growing on it. I spotted a black wolf sleeping near that hill. We gradually saw more wolves, either walking around or bedded down. Eventually, we accounted for all the Druids, except for three of the pups and one adult. Years later I would stand on the hill in that meadow, next to the conifer, and experience the most profound moment I ever had with 21, but that is a story for another time.

After watching the site for three hours, Bill and I started hiking downhill. We got to our car three and a half hours later, tired from our long hike. We were glad we had seen the Druids at their new rendezvous site but were concerned about the missing three pups. Early the next morning, the pack was bedded down at the Chalcedony Creek rendezvous site, still short two gray pups and a black pup. The black pup did a lean-forward urination, indicating it was a male. Since the other black pup was a female, that confirmed he was the male pup who had lingered in the valley and later traveled from the valley floor to the rest of the pack at Opal Creek.

Later in the day the pups played what I called the stealing game. A pup found a bone, bedded down, and chewed on it. Another pup came over, sat down, then suddenly snatched the bone and ran off. In another case a black pup tugged on the tail of a gray who had a bone, probably hoping he would drop the bone to defend himself, but the gray ignored the tail pull and kept a tight hold on his bone. The pups also climbed up on boulders and played king of the mountain. The first pup on top would try to block all the other pups from reaching the summit. If one knocked off that pup, it would control the rock and work to keep the others from that position. I added those two games to the list.

Two days later we found most of the Druids on the south side of the road in Little America. The three pups were still missing. 21 led the wolves west. Soon he passed through the meadow in front of his mother's old den site. I wondered if he took a moment to recall the weeks he had spent there with 8 in the spring of 1997, when the two males had worked so hard to feed and protect those pups. I lost the pack as they went up the far western end of Specimen Ridge. If they went eleven miles east from there, the Druids would end up back at their rendezvous site at Opal Creek.

I had to leave the park for nine days. Bill monitored the wolves while I was away. He saw the Druids at Chalcedony Creek on September 11. They then went out of sight going up the trail to Specimen Ridge, the one Bill and I had taken to look at the Opal Creek rendezvous site. On September 19, they were on Specimen Ridge, above Crystal Creek. There was no sign of the three missing pups, and we wondered if the adults were doing all that traveling in search of them.

I OFTEN THOUGHT that summer about a dark cloud that was hanging over the Druid wolves. The previous December, when I was in Big Bend, the Druids had killed a Rose Creek female yearling, wolf 85. She had been born to 18, 21's sister, in the spring of 1997. I had mistakenly thought with 21 serving as the new Druid alpha male that the bad blood between the two packs might settle down. In mid-July, I was working with Shaney Evans, a Wolf Project biologist who had witnessed that event. She told me it was the Druid females who had killed the Rose female. 21 and 163 were in the group of seven when they chased her, but they were not involved when the five females killed her. While doing research for this book, I contacted Shaney, and she gave me a written account of what she saw that day.

She said the Rose Creek wolves had strayed into the west end of the Druid territory, then moved back west into their own area. The female yearling had lagged behind. The seven Druids came along, saw a wolf in their domain, chased her, then pulled her down and attacked. Shaney then added, "They all ran as a pack when they saw 85, but 21 just stood to the side when they killed her."

Doug was flying that day and he circled over the area during the chase and attack. From his angle, he saw the Druid females leading the chase and takedown. He thought that 21 might have bitten 85 when she was pulled down but did not see him participating when the females killed her. A photo he took of the incident that was later published in Jim Halfpenny's book *Yellowstone Wolves in the Wild* shows a gray wolf that must be 40 and a black female attacking the prone yearling with 21 standing off to the side, looking away from the scene. I later examined the rest of the shots he had taken that day, and in nearly all of them, only 40 is biting 85.

In each picture 21 is away from the attack site and facing the opposite direction.

I had always assumed that 38 had been the instigator of the attacks on the Crystal and Rose packs in the spring of 1996, but I now realized the source of the violence was almost certainly 40. The likelihood she had killed her sister's pups reinforced that thought. She was very aggressive in temperament, while 21 and 42 were more benign. Those two wolves were stuck in a situation they did not know how to resolve. 21 followed a code of behavior that prevented him from using his size and strength against a female, and 42 was too intimidated by the alpha female to stand up to her.

The attack on the Rose Creek yearling meant that the Druids had now killed three adult Rose Creek wolves. I thought about how 8 would view the death of his daughter. Since it happened after 21 joined the Druids, I think 8 would figure 21 was involved. I thought that when the Rose Creek wolves returned to the valley, the likelihood of a second fight between the two adjacent packs seemed higher than ever.

24

The Druid
Yearling Comes
into His Own

I SAW THE DRUIDS five times in late September and never spotted the three missing pups. We reluctantly concluded they had not survived, and we would never know the cause of their loss. I personally felt depressed but accepted the reality that wolves are wild animals and not all of their offspring are destined to survive their first year.

On the last day of September, I found the Druids near the Crystal Creek pen site. I got a report that they had howled from that area earlier and other wolves had howled back at them from Slough Creek to the north. I soon saw the Rose Creek wolves at Slough. They were just a mile or two from the Druids but separated by the park road.

The Rose Creek pack had spent the summer in the high country to the north, and I had not seen them for nearly four months. The founding alpha female was no longer in the pack. Her daughter, wolf 18, had driven her out, just as 40 had driven out her mother. As of 2019, I know of no cases in Yellowstone where a son has driven his father out of the pack. Since I had followed 9 for so many years and knew the tragedies and difficult times she had endured, especially the loss of her original mate, it was depressing to learn that her adult daughter had exiled her and taken over the top position.

But 9 was not a wolf who let setbacks derail her life. She traveled east of the park, settled down in the Shoshone National Forest, and started a new group there known as the Beartooth pack. One of her grandsons, wolf 164, who had been born into the Sheep Mountain pack, joined her. One, then another Rose Creek female also became part of the group. I was amazed at how those other Rose Creek wolves had managed to find their former alpha female so far from the pack's territory, but I later learned that wolves are very good at finding each other over vast distances. The dispersing Rose females had, it seemed, decided to remain loyal to the alpha they had always known rather than switch their allegiance to the daughter who had usurped her. All those wolves were black. Through early 2018, all Beartooth adults and pups have been black. As far as we know, the current members of the Beartooth pack are all descended from 9.

As we got into October, I frequently saw 8 and the Rose Creek wolves at Slough Creek, and they regularly made kills there. 8 was now five and a half years old. With him as its alpha male, the Rose Creek pack had grown to be the largest in Yellowstone. By the end of his life, we think 8 fathered fifty-four pups, as well as raising the eight pups he adopted in 1995.

There were times when the new alpha female, wolf 18, tried to be playful with 8, but he would ignore her and go about his business. If she did a scent mark, he would mark over it and fulfill his duty as the pack's alpha male, but he appeared to have little emotional connection with her. I could not help but have a nonscientific thought: Was he grieving over the exile of 9 by her daughter?

Meanwhile in the Druid pack, 8's adopted son 21 was obviously much closer to 42 than he was to 40. He played with her far more often than he played with her sister. I saw them take turns licking each other's faces. 21 often did play bows to 42 and romped around her. One of his favorite games was getting her to chase him. When he was leading the pack, he repeatedly looked back to check on her. He was more reserved with 40, and they seemed to have what might be called a professional relationship, rather than the playful and flirty one he had with 42. The problem was 40 was the alpha female and that made her the boss of the pack. It did occur to me that the lives of 21 and 42 would be so much easier if something happened to 40.

On October 16, I saw 163 do a raised-leg urination for the first time, at a site the alphas had just marked. He was now in his eighteenth month, about sixteen years old in human terms. In February, he would be capable of breeding a female and could be a father two months after that. The scent marking was a sign of his maturity. That day he led the pack much of the time as they traveled.

There was big news from the October 21 tracking flight. Druid male 104, now two and a half years old, had been seen with the Soda Butte pack in their territory south of Lake

Yellowstone. That was one of the three original 1995 packs from Alberta. The question now was: Had he taken over the alpha male position?

The next day, 163 led the Druids to the west side of Slough Creek. I had not seen the pack there since June 1996 when 8 had defeated 38, the original Druid alpha male. Then the yearling led west, deeper into the territory of the rival pack. It was a dangerous move, for the Rose wolves outnumbered the Druids. But 163 may have had a good reason to take that risk. The other pack had a lot of young adult females, all potential mates. He was following a major travel route used for years by the Rose Creek pack, and he would be getting the scent of those females.

The Druids were back in Lamar the next day. The alphas passed by a bull elk, probably judging him too strong to attempt to kill. But three pups and the three young adult females joined forces and approached the bull. He turned around, faced them defiantly, and charged at the lead wolf, a gray pup. When the bull stopped and ran off, the pups and the females gave chase. Soon five of the six wolves gave up, but the gray pup continued another hundred yards before he too decided it was hopeless.

By late October, 40 was getting more aggressive in her interactions with 42, probably because the breeding season was coming up. I saw 42 approach her sister in an extreme subordinate crouch. Then she sat up in front of the alpha female and licked her face. When 42 got up and walked off, 40 ran over and pounced on her.

21 and 163 had another lengthy play session around that time. The big alpha male wrestled with his son and let him

win. Then 21 ran off with his tail tucked underneath him, pretending to be afraid, and the yearling chased him with his tail raised. On catching his father, the young male pinned him once more. There were two more chases by the yearling, and each one ended with 21 seemingly defeated and pinned. When 21 got up and ran off the next time, 163 stayed in place, so 21 ran back and let his son tackle him. All the other wolves came over and watched the crazy sight of their alpha male acting like the lowest-ranking member of the pack. 21 jumped up, played with the pups for a minute, then went back to 163. The two males took turns chasing each other back and forth. Still in a playful mood, 21 ran circles around the younger wolves. At one point, he wrestled with 105 and pretended she was a difficult opponent for him.

That fall 163 often led the pack and did more frequent scent marking. Since he was out in front, the alpha pair would always find his sites and mark over them. This was a good development for the pack. If a rival group of wolves came into the Druids' territory, they would sniff at every scent mark they found. Now that the yearling was mature enough to join in, the other wolves would correctly figure out that two big males were in the local pack, and that might cause them to leave the area in fear of running into them.

Years later Kira Cassidy, a Wolf Project biologist, analyzed our records of encounters between wolf packs. The three most important factors in determining the winning side were the total count of wolves in each pack, the number of adult males, and the number of pack members over six years old. Having two adult males that both scent marked was significantly better than having just one. Kira found that having one more adult male than their opponents increased that pack's chances of defeating the rival group by 65 percent.

In late October, the Druids returned to the west side of Slough Creek. 163 did a raised-leg urination on a clump of grass, then the alphas also marked his site. The Rose Creek wolves would be sure to find that scent post and get the message that the Druids had visited. I wondered what 8 would think about getting 21's scent. Would he be offended that another alpha male had come into his family's territory and intentionally left signs of his visit? After spending five hours there and doing a lot of scent marking, the Druids returned to their territory.

As the February mating season got closer, the Druid females became more aggressive with each other as they defended their positions in the pack's hierarchy. 40 would pin her sister 42, who would pin 105, the highest ranking of the three younger females. Then 105 would dominate the smallest female, 103, and later the little wolf would pin the lowliest female, 106.

Earlier in 1999, Jennifer Sands, a graduate student under the direction of Professor Scott Creel at Montana State University, had begun a study of stress hormones in Yellowstone wolves. Jennifer wanted to find out if low-ranking pack members experienced high stress rates due to their subordinate status, and how that extra stress affected them. I learned from her work that adrenal glucocorticoid is secreted when animals experience elevated stress levels. That redirects energy from other biological processes, so the animal can devote more resources to dealing with whatever is causing the stress. That redirection can suppress reproduction and lessen resistance to infection and disease.

Jennifer needed to collect scat from identified wolves in the Druid, Rose Creek, and Leopold packs. Other researchers, wolf watchers, and I helped spot wolves defecating, memorized the location, then directed her to the site after the wolf had left. The samples were tested in a lab and the results sent to Jennifer. Unexpectedly, she discovered that the alpha wolves had the highest stress levels. That meant 8 and 21 had more stress in their lives than lower-ranking males. As I thought over that finding, it made sense to me. Both alpha males were responsible for feeding and protecting their families. Each carried the weight of their packs on their shoulders, while subordinate males lived more carefree lives. The same results applied to alpha females. Maybe that partly explained the Druid alpha female's aggressive personality: she was stressed out.

November 12 was the last day I saw 163 with the other Druid wolves or got his signal. He dispersed and likely was looking for a mate.

25

My First Yellowstone Winter

BY THE MIDDLE of November 1999, we were getting early-morning temperatures in the negative range and snow was accumulating on the ground. I bought heavy winter boots and thicker jackets and gloves to help me cope with what was going to be my first really cold winter since 1974/1975 when I lived in Vermont. Back then I had an outdoor job and was used to the subzero weather, having lived through many long New England winters. For the past twenty-five years, however, I had spent nearly all of my winters in desert parks such as Death Valley and Big Bend, and I was worried I might have lost my resistance to cold. I did not know it then, but there would come a day when I was out in minus 54 degree weather. I would need all the extra insulation I could get.

My cabin in Silver Gate was at an elevation of 7,390 feet, so it would be especially cold and snowy there. Some years only seven people stayed in town for the winter. The road from the park headquarters in Mammoth to Cooke City, a few miles east of Silver Gate, was plowed daily, but none of the other park roads to the south were maintained. That worked for me for I wanted to witness what the lives of Druid, Rose Creek, and Leopold wolves were like in the winter months. So far, both in Denali and Yellowstone, I had watched wolves only from May through November, and I was eager to see how they coped with the cold and deep snow. Doug Smith had told me that winter is the preferred time of the year for wolves and that they thrive during that cold season.

One benefit of winter for me was the shorter days. I could sleep in until 5:30 a.m., leave the cabin an hour later, and get out to Lamar Valley in plenty of time for first light, a little after 7:00. After rising by 3:15 a.m. in June and July, that extra time to sleep in was luxurious. It began to get dark around 5:00 p.m., so I got home long before my typical summer return time, which was usually after 9:30 p.m.

On November 15, the Wolf Project's Winter Study started up. Doug Smith began the study in the fall of 1995, based on a similar study he had worked on at Isle Royale that went all the way back to 1958. The Winter Study was divided into two thirty-day observation periods, one from mid-November to mid-December and another one in March.

Doug brought in extra people for the Winter Study, usually wildlife biology graduates with field experience. There were three crews of two people, each studying separate packs: Druid, Rose Creek, and Leopold. The study crews were out from first to last light and recorded everything the wolves did, minute by minute. The study took particular note of hunting behavior and documented all chases, both failed and successful. When their pack made a kill, the crew would later visit the site and conduct a necropsy on the carcass to determine the animal's age and condition.

Weather permitting, a tracking flight also monitored the packs every day during the study period. The plane looked for kills that might be out of sight to the ground crews. All those carcasses would be added up at the end of each observation period so kill rates could be compared from year to year, and the prey species and the age and condition of each animal could be tallied. Over many years of collecting data, we found that an adult wolf needs 1.4 to 2.2 adult elk carcasses per month in the winter to stay in good health. Those elk could be a combination of ones the wolf killed and ones that died of natural causes. One adult bison could substitute for perhaps three elk.

Tom Zieber and Rob Buchwald comprised the Druid crew, and Shaney Evans and Dan Stahler were the Rose Creek team. One day I was with Shaney and Dan at Slough Creek, watching the Rose wolves. 8 and the five wolves with him howled. We heard the Druids howling back from the southeast. Soon the Druids came into sight to the south of Slough Creek. They had moved toward the sound of the other pack's howls. But it was just the alpha pair. The other Druids were lingering at a carcass ten miles to the east. The Druid alphas veered uphill to the south, traveling away from the Rose Creek howls. We think wolves can estimate how many animals are howling at them from a rival pack. Since 21 and 40 were outnumbered six to two, avoiding the Rose Creek wolves was a smart move. As 21 led south, I also wondered whether he moved off because he did not want to get into a confrontation with 8.

IN LATE NOVEMBER, the temperatures dropped further into the minus range. One day I saw the Druids walk across an ice-covered pond. A pup grabbed a piece of ice and romped around with it. He tossed it in the air, then got another piece and threw that up as well.

That day, eight of the Druids were south of the Lamar River, just east of Tower Junction. Meanwhile, 8 and twelve other Rose wolves were slightly north of the river. Near-meetings of the two neighboring packs were getting more common. We were not sure if 21 knew the Rose pack was so close. The two packs had not howled at each other, but if the wind was right, he could have gotten the scent of the other wolves. For whatever reason, he led his pack south, in the opposite direction, and that prevented a potential clash between the two groups.

One morning I saw 21 and a pup chasing a herd of about fifty elk. The pup was out in front, running all out. 21 seemed to deliberately stay behind the pup, perhaps worried the elk might turn back and trample the little wolf. The herd split up and elk ran in different directions. Confused about which ones to chase, the pup stopped in the middle of the melee and 21 stopped as well. Then the pup ran after one subgroup and 21 resumed following him. A minute later the pup was gaining on one cow. The elk stopped, turned, and confronted him. The pup ended his pursuit and just stood there, not knowing what to do next. 21, deciding the cow was not a suitable target, trotted off and the pup followed.

The pup soon got distracted and fell so far behind 21 that he could no longer see him, but, already experienced at following scent trails, the pup expertly stayed on his father's exact route. He went out of sight to the east, and I had to drive several miles to get to a viewpoint where I could wait for him to reappear. When I spotted the pup again, he was running after seventy-five elk. No other wolves were visible. The pup soon was in the middle of that herd. Once again, the herd split into several subgroups and the pup went after the nearest one. Those animals split again. The lone pup was now chasing a cow and her calf. The pup concentrated on the calf, something more his size than its 450-pound mother. Going as fast as he could, the pup was closing in. It was just thirty yards behind its intended target. Then the calf kicked into high gear, leaving the wolf far behind.

The calf reached the herd and ran into the middle of scores of elk. A cow ran out at the pup, and he had to turn around and flee from her. The other elk spooked at the scent of the wolf, not knowing it was just a pup, and ran off. The pup turned around and resumed his chase. Two big bull elk stopped and confronted him. Undeterred, the little wolf charged at them. One of the bulls ran at him. The pup gave up on the elk. He trotted off, looked around for the other Druids, found their scent trail, and followed it east. I later saw that pup invent a new game. After catching a vole, he repeatedly tossed it in the air, then leaped up and tried to bat it with a front paw as it dropped to the ground. I added baseball to the list of games pups play.

Tom and Rob told me about an encounter they had just seen between the Druids and the Crystal Creek wolves. It had taken place a few miles south of Specimen Ridge, in an area the Crystal wolves would have considered their territory. The Druids had seen one wolf from the rival pack and chased it. That wolf ran south to the rest of the Crystal wolves. They spotted the Druids and charged at them. Surprised by the appearance of additional wolves, the outnumbered Druids ran back to the north, toward their territory.

40 and the black pup lagged behind the other Druids as the Crystal wolves raced after them. 21 was ahead of them, staying close to the two gray pups, acting as their bodyguard. He led them down into the Yellowstone River corridor, and the other Druids followed his route. The Crystal pack also ran into that area. Three days later, we spotted the Druids and noticed the black pup was missing. We never saw him again and concluded the Crystal wolves had probably caught up with the pup and killed him.

Earlier I wrote that when 21 joined the Druids he inherited two feuds: one with the Crystal Creek pack and the other with the Rose wolves. The Druids had killed the Crystal Creek alpha male in the spring of 1996. His mate from that time, alpha female 5, likely was with the other Crystal wolves when they caught and killed that Druid pup. Seen from a human point of view, 21 and his son had nothing to do with the killing of the Crystal Creek alpha male. But the rival pack's vengeance was visited on 21's offspring now that he was the Druid alpha. That meant that with the year about over, only two of the six Druid pups were still alive, a survival rate of 33 percent. The Wolf Project keeps records of pup survival from birth in April to the end of the year, and from 1995 to 2017 the average survival rate was 73 percent. The previous year, 1998, one of two Druid pups had survived. The pack was not doing well in raising their pups.

edge of the opening in the ice, then slipped into water partway up his chest. The calf, which was taller than the wolf, charged through the water at 8 and tried to stomp down on him with its front hooves. The wolf dodged the blow. The calf made a second charge but missed again. On its third attack, 8 did not get away quickly enough and was struck in the head, a blow that must have stunned him and might even have given him a concussion.

That last kick, however, threw the calf off balance and it fell down into the freezing water. Ignoring the punishing blow to his head, 8 charged. The calf jumped to its feet and ran. In its desperation, it managed to climb out onto the ice, where it immediately slipped and fell. Jumping out of the pool, 8 ran to it, but the calf got up an instant before he arrived and plunged back into the creek. The wolf jumped back in as well and chased the calf through the water.

The black wolf watched from the other side of the pool but made no attempt to help. I would have hesitated to enter the freezing water as well. Trying to estimate the air temperature, I figured it was around 15 degrees Fahrenheit. I admired 8's grit for twice jumping into the creek on a very cold December day and continuing to go after the elk after getting a hard kick to his skull. Getting kicked and soaked in cold water were things he had to endure if he was to feed his family. It was part of his job.

More pack members arrived, but all stayed on the ice and, like the first black, just watched as 8 battled the elk by himself. They saw the calf chase 8 through the pool and vigorously kick out at him. The wolf just barely dodged those strikes. Then he chased the calf through the water. At first it ran, then it stood its ground and lashed out at the wolf with its front legs as they stood face to face.

I have a vivid memory of the calf rearing up and coming down on 8's back with both front hooves. Its weight drove the wolf underwater. If it could pin him there long enough he would drown, but 8 squirmed out from under the elk and resurfaced. I winced as I saw that now he was really hurting. If this had been a mixed martial arts fight, the referee would have stopped the match, fearing 8 was too injured to defend himself.

At that moment, when his father needed help the most, the first black jumped into the water. It teamed up with 8, and they both chased the calf through the pool. Five other black wolves stood nearby on the ice but did not help. The calf kicked out at the black and knocked it down. The young wolf got right up and rejoined 8 in chasing the calf. Now exhausted, the calf was slowing down. That was what 8 had been waiting for. Despite being kicked in the head and stomped on the back, and likely hypothermic, he ran through the water, leaped up, and grabbed the calf's throat. As he held on, the black helped by getting a holding bite on a back leg. It got kicked in the head by the other hind leg and lost its grip but came right back and grabbed the leg again. The young black was just as determined as 8 to finish the job, regardless of the cost.

The other five blacks still did not jump in to help. I now saw that all of them were pups. 18, the alpha female, ran in and joined the five but did not help 8 and the black with him. The six wolves looked like spectators at the Roman Colosseum watching gladiators fighting to the death.

As the calf struggled, the black lost its hold on that back leg once more, but 8 still held on to its throat. His position made it impossible for the calf to strike out at him with a front hoof. The black ran through the water to 8 and bit the calf's head. Both wolves then worked together to drag the struggling elk out of the water and onto the ice shelf. At that moment, 8 let go of its throat.

The five black pups ran to the calf, which was lying on its side. I wondered if 8 intentionally released the calf to give his pups the experience of finishing it off. But the calf's frantic attempts to stand up on the slippery ice frightened the pups, and they backed off. After a few moments of hesitation, they ran back and helped their father and the first black kill the calf.

I could see that 8 was in bad shape. In a few months he would be six years old, the equivalent of a man in his late forties, and all the wear and tear on his body was adding up. We eventually learned that 8 was in far worse physical condition than we realized, due to years of fights with elk such as this one.

While writing this book, I spoke to Jim Halfpenny about 8's skull. Sue Ware of the Denver Museum of Nature and Science and Jim had examined it after 8's death. Jim told me two canine teeth were missing, one was broken off, and the fourth one was worn. A wolf's four canines are the long, sharp front teeth used to bite and kill prey. Several other teeth were missing or broken. There were a lot of abscesses, meaning infections, in his jaws. Jim said that there would have been a bad smell from those areas. 8 would have learned how to detect the scent of sickness and infection in prey animals and would know that he was now giving off that scent himself.

Jim loaned me two photos of 8's skull, and I saw that his jawbones were honeycombed with gaping holes from those infections. The front of his lower jaw looked more like a sponge than bone. The pain he must have endured from those abscesses would be unimaginable.

The spacing between the two broken teeth on his right side, the upper canine and upper premolar, was about the same width as an elk hoof. Based on that, Jim felt there was a good chance 8 had been kicked squarely on the right side of the jaw by an elk. There were signs of healing, which indicated that injury had happened well before his death. On thinking about Jim's comments, I realized that I might have witnessed that kick to his head during 8's fight with the calf in the creek.

Many years later, when I was reading about how so many NFL players suffered from brain damage after receiving repeated blows to their heads, I wondered if 8 had the same syndrome. A study of 111 brains of deceased professional football players found 110 had evidence of degenerative brain disease (chronic traumatic encephalopathy or CTE), due to those hits. All those players wore helmets. 8 had nothing protecting his brain.

I tried to calculate how many successful hunts 8 had been on during his fifty-six-month tenure as alpha male and figured it easily had to be several hundred. Consider the impact of the punches Muhammad Ali took to his head during his sixty-one professional fights. 8 had way more fights, many of them fights to the death with animals far bigger and stronger than he was. All the blows he received made him old and disabled before his time, like they did Ali. Ali retired, but 8 did not have that option. Whatever injuries he endured, he had to continue to go out on hunts and expose himself to more blows. He soldiered on, fueled solely by willpower. But the reality was he could not continue like this much longer.

I thought about a quote from Ali that poignantly summed up his life, as well as 8's: "What I suffered physically was worth what I accomplished in life. A man who is not courageous enough to take risks will never accomplish anything."

IN MID-DECEMBER, I got the Druids' signals two days in a row from the area behind the Yellowstone Institute. Mountain lion researcher Toni Ruth came by and told me she had radio collars on a family of cougars and was getting their signals from the same direction. It was a mother who was raising four six-month-old young lions. Wolves and cougars, like dogs and cats, are natural enemies and we wondered what was going on up there.

Two days later, the wolves left the area and traveled east. Later that day, I went to the Institute to check on the lion signals. The collars on two of the young cougars were on mortality mode. I got normal signals from the other three collars. I contacted Toni and she came over with her research crew. We trudged north through the deep snow for a mile and a half. On the way, Toni told me she had first picked up the mother lion's signal from this area on December 8 and thought the cougar had made a kill that day, probably an elk. I had detected the Druid signals from that direction on December 13 and had them there the two following days.

We arrived at the site and found the remains of an adult cow elk. Nearby was the carcass of one of the young lions, under a few inches of snow. Toni examined the tree closest to the elk and saw claw marks on its trunk. Several nearby trees also had lion claw marks on them. That suggested that when the Druids arrived, the lions climbed up those trees. After more searching Toni found a second dead lion, the sibling of the first one. Tracks in the snow showed it had been chased and killed by the wolves. The same thing must have happened to the other lion. Both must have jumped down from their trees, naively thinking they could go back to the elk carcass without a response from the wolves. Each of them was about 40 pounds, similar in size to a big coyote.

Five days after we found the remains of the two young lions, I spotted the Druids near an elk they had just killed. Toni came over and got signals from the mother lion up in that area, but both of her two surviving young were on mortality mode. It looked like the lions and wolves had fought over the carcass, and the Druids had won the battle.

The next month, our tracking flight got a mortality signal from wolf 163 east of the park. He had recently dispersed from the Druids and was probably looking for a mate. A helicopter flew in a few days later and Kerry Murphy, a wildlife biologist who had joined the Wolf Project in March of 1998, examined the site. Kerry had to dig the wolf out from under two feet of snow. The remains of a bighorn ewe were also in the snow near him. The ewe had been consumed by one or more predators. There were lion tracks and scat nearby. Toni told me that a 130-pound male cougar lived in that area.

Due to the time elapsed and condition of his remains, Kerry could not determine the cause of 163's death, but the evidence at the site caused me to consider the lion the likely suspect. I later heard of a man who was hunting in western Montana. He saw a wolf standing under a tree and paused to watch it. Suddenly a big cougar jumped down from the tree, grabbed the wolf by the head, and crunched its skull, killing it instantly. Perhaps something like that had happened to 163.

A few years later, I found the partial remains of one of 21's granddaughters near a bull elk carcass close to the road. Her head, three legs, several ribs, and a lot of clumps of fur were at the site. The rest of her body was missing, apparently eaten. I investigated the incident and got a report that a wolf had been seen heading to the carcass three days earlier. That night, a Wolf Project volunteer had seen a big mountain lion running across the road toward the site. We went back to where we had found the wolf and spotted a lion track. We concluded that the lion had probably found the wolf at the carcass and then killed and eaten her. That incident helped me understand why the Druids had killed those four young lions. 21 regarded them as a threat to his family. The subsequent death of his granddaughter proved him right.

AS WE GOT into late December, deep snowfall made travel difficult for the Druids. 21 usually led the pack, using his prodigious strength to break trail for the others. To conserve energy, when possible he used existing trails made by bison or elk.

Tension was building between the Druid and Rose wolves. Both packs howled at each other frequently, the Druids from

Lamar Valley and the Rose pack from Slough Creek. Sooner or later, the two adjacent packs would meet up.

On Christmas morning, I found the Rose Creek wolves with a new elk carcass west of Slough Creek. The pack had already fed and was bedded down away from the elk. Several coyotes were sneaking in to the site, stealing meat, and running off with it. 8 and two blacks jumped up and ran over. The snow at the kill site was several feet thick. The wolves had tramped down a pit around the site that was deep enough to keep a coyote from seeing the approaching wolves.

The three Rose wolves charged into that pit. I got glimpses of a young coyote, possibly a pup, snapping at the wolves and of the wolves attacking it. Soon the coyote dropped out of sight. I saw wolves biting at something at the bottom of the pit. 8 stopped his attack and the other two wolves followed his example. I was sure the thief must be dead. A raven landed, anticipating a meal of coyote meat. The wolves left, done with their mission.

I followed the wolves as they trotted to their bedding site, then swung my scope back to the carcass. A few minutes later, the head of the coyote peeked out from the snow pit. It looked around, saw that the wolves were far off, and ran in the opposite direction. I saw no obvious serious injuries on it. 8 and his family had killed that elk and had right of ownership. The coyote was stealing meat that belonged to the pack. 8 would have been justified in killing that coyote, but he just beat it up, then let it go to live another day. His actions reminded me of how three and a half years earlier, in this same area, he had defeated 38, the big Druid alpha male, then let him go.

On Christmas afternoon I watched the Druids in Lamar Valley. They were bedded down near their latest kill. One of the young females led off to the west and 21 immediately followed. All the other Druids fell in line behind him. The pack stopped to socialize, and 21 went to 42. In a playful mood, he rolled on his back under 42, pretending he was subordinate to her. Another young female and one of the two surviving gray pups saw 21 on the ground and ran over. Anyone coming on the scene without any prior knowledge of the pack would conclude that the three wolves were clearly dominant to the one on his back. Once again, 21 was playing a pretending game. The two adult females walked off, but the pup continued to stand over his father, acting like he had just defeated him in a wrestling match.

Then 21 jumped up and romped off, still in a playful mood. His tail was tucked down over his rear end as the pup chased him. Running over to 42, he wagged his tail and romped alongside her, like a pup greeting its mother. The next moment he fell over on his side, the way a comedian might do a pratfall for laughs. There was only one word to describe the antics of this father wolf as he played with his family on that Christmas day: goofy.

I have now watched and studied wolves for forty years in Alaska, Montana, and Wyoming, and I have never seen another alpha male behave like that. But when 21 stood up, anyone could see him as he really was: the toughest, most powerful wolf in the park. If a Marvel Comics artist were to draw an idealized superhero wolf, a new member of the Avengers, the portrait would look like 21 did at that moment.

The next day, 8 and the Rose Creek wolves were on the

27

The Battle of Specimen Ridge

O N JANUARY 12, 2000, the fifth anniversary of the re-introduction of wolves to Yellowstone, I saw both the Druid and Rose Creek packs. 8 was the only wolf in either pack who had arrived in the park on that date in 1995. The other five surviving members of that original batch of fourteen wolves were now alphas in their respective packs: 2 and 7 were the Leopold alpha pair, 5 was the Crystal Creek pack alpha female, 9 had left the park to become the alpha female in the Beartooth pack, and 14 was the Soda Butte pack's alpha female. For all of us that lived through those five years, including the wolves, it had been a time of intense drama and experience.

Ten days later, I got signals from the Rose Creek wolves on top of Specimen Ridge in Druid territory. The Druids were to the east at their den site near the Footbridge and Hitching Post parking lots, and I heard them howling. That should have caused the Rose pack to leave, but the following day, they were still up on the ridge. On January 24, 40 led a group of seven Druid wolves to the lower part of Specimen Ridge, indicating that she knew where the Rose Creek wolves were. The rival pack had been in Druid territory for three days now, and I sensed that a major confrontation, a long time in the making, was finally about to happen.

At 11:55 a.m. that day, the Druids had a group howl that functioned as a claim to their territory. I heard distant answering howls from high on the ridge that had to be from the Rose wolves. The Druids immediately stopped howling, then ran in the direction of the challenging howls. I saw the Druids off and on as they ran uphill through forests and meadows. They needed to climb over a thousand feet to reach the source of those howls. Running that far uphill would tire them out and put them at a disadvantage if they had to fight the other wolves.

21 led the pack. I thought about how he and 40 had tried to get their pups out of harm's way the previous month by running away from the Crystal Creek wolves and how that strategy had failed to save the black pup. On that day, the Druids had been out of their territory. Perhaps the alphas had been unsure of themselves in unfamiliar country. But now they were defending their homeland, and as I looked at 21, I saw that he was deadly serious. He was now at his physical peak, able to defeat any opponent.

The distant howling continued. I then saw two black wolves standing on the crest of Specimen Ridge, staring in the direction of the area where the Druids had first howled. Suddenly, the pair ran off along the ridge. They had probably seen the enemy pack charging up at them. They stopped and looked back. I swung my scope to the Druids and saw them racing toward the spot where the two blacks had last howled. The seven Druids soon were on their scent trail. They spotted the blacks and chased them, then turned back, searching for the main group of invading wolves.

A few moments later I saw the rest of the Rose Creek pack. 8 and seven others were downhill from the Druids, running along their scent trail. The Druids did not know that the other wolves were behind them and rapidly closing in. 8 was leading the Rose group, his tail held high. Like 21, his job as alpha male was to defend his family. Nothing was going to stop him from fulfilling that duty, even if he had to fight 21. But since two of his canine teeth were missing and a third one broken off, any bites he inflicted on a rival would be ineffectual. Considering his deteriorating physical condition; the near uselessness of his only offensive weapon; and the size, strength, and fighting ability of his opponent, 8's determination to battle 21 was the bravest thing I had ever seen.

That was the moment I realized what was about to happen. 21 would soon see the other wolves and charge at them. He would single out their alpha male and confront him. I could not imagine 8 backing down or running away when it came to protecting his family, so the two males, a father and son, would fight, each thinking he was doing the right thing. From 8's point of view, he was charging at the pack that had killed three of his family members. 21, for his part, had lost four of his six pups that year and had to stop these wolves from killing the last two. It would be a fight 21 could not lose, and a fight 8 could not win.

8 and the seven wolves with him were now running directly at the Druids. I picked up the seven Druids, who had changed direction and were now running right toward the Rose wolves. 21 was out in front of his group, and 8 was leading his pack. Both alphas were running straight at each other, like two medieval knights on horseback with their lances aimed at each other's chest. Did 21 know this was 8, the wolf who had adopted and raised him? Once he got a good look at the lead gray wolf in the opposing pack, he would recognize him. He would then have only a few moments to decide what to do. It would come down to this: What was more important to 21, protecting his family or his loyalty to his father?

A thought flashed into my mind. There was a way 21 could deal with this conflict without killing his father. He could engage 8, use his superior strength to take him down, then let him go, just as he had seen 8 do years earlier with 38. But I immediately saw the fatal flaw in that strategy. 40 was right behind 21. If 21 pinned 8, she would run in a moment later, before 21 could let 8 go, and attack him, like when she had killed the Rose Creek female a year earlier. Her super-aggressive personality would trigger an all-out attack when the other Druids ran in and joined her. 8 would not survive such an assault. Those were my thoughts. I had no idea what was going on in 21's mind, especially since he surely felt that his family was threatened. There was no obvious way for 21 to protect his pack and somehow give 8 a chance to survive.

Now the two lead males were just a few yards apart and rushing forward. A head-on collision was about to happen. I tensed up, expecting the worst.

An instant later, 21 shot right past 8 without touching him. Less agile, 8 continued to run forward, then realized what had happened, turned around, and ran after 21, proba- bly thinking the Druid alpha was running away from a fight. All the other Druids, confused at what their alpha male had just done, followed his example and ran through the Rose wolves. Both packs were now disorganized. Wolves chased wolves, got distracted, and chased other wolves, but I saw no fighting. I found 21 and saw that four Rose wolves were pursuing him. He sped up and led them away from the other Druids.

After several minutes of running around in confusion, the Rose Creek wolves ended up east of 21. They came together and had a rally, jumping on each other in excitement. 21 howled at them and got their attention. Howls came from west of him, from the Druids. 21 was positioned between his pack and the Rose wolves. At that moment he ran forward, toward the rival pack. But 21 was not on the attack. He was running to the place where I had just seen one of the two pups remaining from this year's litter, who had gotten sep- arated from the Druid adults. He was focused on reuniting his family.

Six Rose wolves were now on top of a steep snow cor- nice, uphill from where I had last seen the Druid pup. 8 and a black were below the cornice. He, too, wanted to keep his pack together after the chaotic confrontation, and he con- centrated on going around the cornice to the upper wolves, rather than going after 21 or the Druid pup. I heard that pup howling and spotted him downhill from the rival wolves. 21 was last seen going that way.

The Rose wolves regrouped and ran back and forth but failed to find any Druids. About an hour after 21 ran past 8, I saw 8 and his family bed down, exhausted. At that time,

40 was the only Druid wolf in sight. She was down near the Lamar River, traveling east and howling constantly as she tried to get the other Druids to come to her.

The next morning, I got the Druid signals from their den forest. I later saw the Rose wolves miles west of there, back in their territory. There were ten, the same total as I had yesterday: eight in the main group and the two blacks I had seen above them. I returned to Lamar Valley and spotted all eight Druids, including their two pups. None of the wolves from either pack was missing or appeared injured. After many hours of not knowing if any wolves had been killed or hurt, I could finally relax.

IN LATE JUNE 2000, after 18, the new Rose Creek alpha female, had given birth to another litter of his pups, 8 went on a hunt way up Slough Creek. The tracking plane later flew over that area and got a mortality signal from his collar. Doug Smith and Kerry Murphy rode up Slough Creek and found his body wedged under a log in a shallow, slow-moving section of the creek. The details of that location argued against his death occurring there. Both men noticed there was blood coming out of the wolf's nose, a sign he had been injured.

Doug figured that 8 had likely chased an elk into the water upstream from that site, fought with it there, and got kicked in the head. The blow might have killed him outright or more likely stunned him so severely that he drowned, but either way, the cause of his death was the elk. The current then carried his body downstream to where he was found. Doug's theory fit in well with Jim Halfpenny's assessment that the earlier damage to 8's jaw was probably caused by an elk kicking him in the head. He managed to survive that blow but not the final one.

Considering that most wolves are killed in fights with other packs or are shot or trapped by humans, 8 had a good death. Dying in combat was an honorable ending to his life. To his last breath, he was serving his family.

After writing those words, I thought about what 8's last few moments in the creek might have been like. Once he realized he was not going to survive, maybe he stopped struggling and prepared to die, the way dogs often relax at the very end of their time. I would like to think that as all his terrible pain was fading away, his last thought was one of gratitude for the life that was given him.

IT IS NOW nineteen years after that incident involving the two alpha males on Specimen Ridge. Since then I have thought about the event thousands of times and this is what I have concluded. Before he left the Rose Creek pack, 21 was always respectful to 8 and willing to be subordinate to the wolf who had raised him, just as he would have been to his biological father if wolf 10 had not been shot and killed. On that January day, I think 21 tried to avoid a battle between the two packs by pretending to run away from 8. When the entire Rose Creek pack chased 21, that allowed the younger Druids, including his two pups, to get safely away.

21 ran away, just like he pretended to run away from his yearling son 163 when they played together, just like Kintla pretended to run from me around that table, just like my father pretended to lose to me in that long-ago

wrestling match. 21 turned what could have been a deadly fight between father and son into a game of catch me if you can, knowing he could outrun all the other wolves.

The younger Rose Creek wolves saw this huge alpha male from a rival pack charging at them and saw their alpha male bravely run directly at him, despite all his accumulated injuries and broken teeth. It must have looked to them like the other male saw 8, feared him, and ran off. Then they witnessed their father chase him far away. To them, it was clear what happened: 8 had won the match.

I think 21 did what he did out of respect for what his father had done for him. It would be like a Samurai warrior having a misunderstanding with his much older master and refusing to fight him, out of respect and gratitude, not caring if any witnesses might think him a coward.

21 inherited his size and strength from his sire but was raised, trained, and mentored by his adoptive father. 8 showed his adopted son how an alpha male and father wolf should behave. Experts say dogs learn best by watching and imitating other dogs. That is what 21 did: he watched and imitated 8, the only father figure he ever had.

As the years went by, 21 never lost a real fight with another male. He was the undisputed, undefeated heavyweight champion. I once saw him fight six rival wolves by himself and still win. All that proved that 21 had no fear of getting into fights and reinforced my thinking about why he ran past 8 that day. 21 continued the tradition he had learned from 8 when he saw his father defeat the original Druid alpha male, wolf 38, and let him go. As far as I know, 21 never killed a defeated rival.

At that time, 21 still had half of his life yet to live. He became the most famous male wolf in Yellowstone and perhaps in the world. Due to the many Bob Landis Yellowstone wolf documentaries on the PBS *Nature* series and other television channels, 21 was a living legend. Some suggested that he was the greatest wolf that ever lived. I would agree with that. But if somehow 21 had any awareness of that acclaim, I think he would have rejected the title. In his mind, there was a wolf greater than him. It was the only male he would defer to. It was the wolf who adopted and raised him. For 21, the greatest of all wolves was 8.

EPILOGUE

Nemesis

I N THE NEXT few months, there would be a violent rebellion in the Druid Peak pack, and 21 would have to deal with its dire consequences. After that, a young male wolf would arrive in Lamar Valley and create endless chaos for 21 and his family. It would be his nephew, a wolf with a personality the exact opposite of 21's. He would become 21's nemesis. The story of 21, his relatives, and his descendants will be continued in the next two books in the Yellowstone wolf trilogy.

AFTERWORD

NO ONE HAS spent more time observing wolves in the wild than Rick McIntyre. No one. Take a moment to think about this. The person who wrote the book you have in your hands has dedicated more hours, days, and years to documenting the lives of wolves than anyone who has ever lived. This alone is a remarkable feat. Then add Rick's ability to spin what he has seen into a compelling narrative. If observation is one of his great strengths, storytelling is another.

Here are a few specifics to make this astonishing work come alive for you. Rick has spent forty years of his life watching wolves in America's national parks. At one point in his twenty-five years in Yellowstone, from June 2000 to August 2015, he went out for 6,175 consecutive days. That's over fifteen years of daily trips in search of wolves. Only a serious health condition broke this streak. (He is okay now, and the stress of his "streak" is also over.) Rick likes to tell me that his record was better than Cal Ripken's streak of consecutive games played in Major League Baseball. (Rick doubled Ripken's record, plus Ripken had an off-season to recover.) During one particularly productive period, Rick saw wolves 892 days in a row. At the time I am writing this, he has accumulated 99,937 wolf sightings, and he will keep at it until he breaks 100,000. The journal he has kept every day fills 12,000 pages. Rick is a man of detail and records. I can't imagine anyone ever being more dedicated or meticulous.

I have made the study of wolves my life, specifically the science of them. My research led to the genesis of the Yellowstone Wolf Project back in the mid-1990s. Adding Rick's observations and insights to the project's scientific research has made the study of wolves in Yellowstone perhaps the most significant study on wolves ever conducted. Our scientific methods, for the most part, are standard, but add in those 99,937 wolf sightings, along with thousands of hours of observation time, and you start to get what I'm talking about. I can't count how many times I've asked Rick, "What do you think is going on here?" or "How many times have you seen that?" Most importantly, this combined approach has opened a window into wolf life, the furthest we have come in understanding how wolves think. This is the single most difficult, some even say impossible, accomplishment for animal science. If you know what an animal is thinking, then you can know the animal. We always fail. With the help of our scientific research, Rick has come as close as anyone to this phenomenal feat: understanding the thoughts of a wolf.

Rick's other great contribution to Yellowstone and the wolves is himself. Over the years, Rick has become more than just a treasure trove of information; with his quick wit and eccentricity, he has become a sought-after storyteller. Everybody wants to meet Rick and hear what he has to say. He has a special way of telling what he knows, as this book attests. He also has *always* had patience for people, and he has *always* been willing to stand and wait for one more person coming in to see or hear about a wolf. Thanks to this generosity, Rick has become a global ambassador for wolves. He has been especially supportive of young people, spotting them in the field and seeking them out to help, talk to, and uplift. I cannot recall a single time when he turned down a student who wanted to shadow him or do a project with him. He does all the media interviews, too—everyone wants his take—but his true passion, after the wolves, is helping young people.

This book is a small slice of Rick and his stories and his insights into wolf life. Treasure it. Know that what you have just read is one of the most personal and in-depth insights into wolves ever produced. We will likely never know where Rick's passion came from. We are just lucky to have seen it and benefited from it. None more so than the wolves themselves. The beauty of the wild wolves of Yellowstone is that they are other and apart from us. It is not within their power to express gratitude in human terms, and maybe they wouldn't even want to, but if it was and if they did, there is no doubt in my mind that they would express their thanks to Rick.

DOUGLAS W. SMITH, senior wildlife biologist and project leader for the Yellowstone Gray Wolf Restoration Project, January 2019

As this book was in typesetting, Rick reached his goal of 100,000 wolf sightings on January 27, 2019. To be precise, as Rick likes to be, he reached 100,001 by the end of the day, and even though he is now officially retired, he is showing no signs of slowing down.

AUTHOR'S NOTE

INSPIRED BY HOW members of a wolf pack support each other in time of need and by all the kind people who have helped me over the years, I will be donating proceeds from my wolf books to Yellowstone National Park and to non-profit organizations such as the Make-A-Wish Foundation of America and the American Red Cross. Readers who are interested in helping support wolf research and wolf education in Yellowstone can go to the Yellowstone Forever website, www.yellowstone.org/wolf-project, to make a donation. Yellowstone Forever is the official nonprofit partner of Yellowstone National Park and helps fund the park's Wolf Project, the National Park Service operation I used to work for.

ACKNOWLEDGMENTS

I FIRST WANT TO thank my editor, Jane Billinghurst, for working way beyond the call of duty to make my original manuscript far more readable and polished than it would have been without her help. Thanks also to Rob Sanders at Greystone Books for accepting my proposal for this book and the two others in the trilogy. Everyone else at Greystone was very supportive and encouraging and thanks to all of them.

My good friends Laurie Lyman and Wendy Busch read the first draft of the book and gave me very helpful comments and suggestions on how I could improve it.

There were many National Park Service staff members, wildlife researchers, and filmmakers here in Yellowstone who advised me on their work and experiences with wolves. Many of them gave me valuable suggestions on sections of the book that related to their field of studies. Here are the ones that were especially helpful: Norm Bishop, Lizzie Cato, Shaney Evans, Anne Foster, Jim Halfpenny, Mark Johnson, Bob Landis, Debbie Lineweaver, Kerry Murphy, Carter Niemeyer, Ray Paunovich, Jim Peaco, Rolf Peterson, Jack Rabe, Toni Ruth, Dan Stahler, Erin Stahler, Jeremy Sunder-Raj, Linda Thurston, Chris Wilmers, and Jason Wilson. I want to give special thanks to Kira Cassidy, who drew and illustrated the map for the book. There have been scores of volunteers that have worked for the Wolf Project over the years and every one of them was very helpful to me. The wolf reintroduction might never have happened without the support of three critical National Park Service leaders: National Park Service director William Penn Mott and Yellowstone superintendents Bob Barbee and Mike Finley.

Special appreciation goes to Doug Smith, the lead biologist for Yellowstone National Park's Wolf Project. Doug is a unique PhD-level scientist who can relate what he has been learning about wolves and the natural world to regular people in ways that not only educate, but more importantly inspire them. Despite Doug's heavy workload and obligations to his family, he took the time to read my manuscript and suggested valuable changes and additions. I carried out every one of them and they made this book much better than it would have been without Doug's generous involvement. I especially thank him for his kind words in the afterword.

There have been hundreds of wolf watchers who have greatly aided me over the years. On many occasions when I was trying to find wolves another person would spot them first and graciously point them out to me. I would also like to thank the vast numbers of Yellowstone visitors that have been kind and friendly to me over my many years here. There is something about being in the park that seems to make people positive and sharing. Thank you to everyone that I have met over the years. I could not have done this without all of you. I regard this book as a joint effort.

REFERENCES
AND SUGGESTED
RESOURCES

Direct Quotes

"That which does not kill us..." Friedrich Nietzsche. 1977. *Twilight of the Idols*. Translated by R.J. Hollingdale. Harmondsworth, UK: Penguin. Originally published in German in 1889.

"Being a hero means..." Dwayne Johnson. 2013. *The Hero* reality show, TNT.

"Even the smallest person can change the course of the future." J.R.R. Tolkien. 1954. *The Fellowship of the Ring*. London: George Allen and Unwin.

"Today is a good day to die..." Attributed to Low Dog, Oglala Sioux Chief who fought with Sitting Bull at Little Bighorn.

"A boy wants to be just like his dad..." Kevin Von Erich. 2014. Interview, *Snap Judgment* podcast, January 24, 2014.

"Yellowstone is home..." Yellowstone Mission Statement, *Yellowstone Resources and Issues Handbook, 2017*, page 22.

"We choose to go to the moon..." President John F. Kennedy. 1962. Speech, September 12, 1962, Houston.

"What I suffered physically..." Muhammad Ali. 1984. News conference, October 28, 1984, Houston.

Research Material

Cassidy, Kira, Daniel MacNulty, Daniel Stahler, Douglas Smith, and L. David Mech. 2015. "Group composition effects on aggressive interpack interactions of gray wolves in Yellowstone National Park." *Behavioral Ecology* 26: 1352–1360.

Cooper, H.W. 1963. *Range and site condition survey, northern Yellowstone elk range, Yellowstone National Park*. Bozeman, MT: USDA Soil Conservation Service.

Duffield, J., C. Neher, and D. Patterson. 2006. "Wolves and people in Yellowstone: Impacts on the regional economy." Missoula, MT: University of Montana.

Halfpenny, James. 2003. *Yellowstone Wolves in the Wild*. Helena, MT: Riverbend Publishing.

Hart, B.L., and K.L. Powell. 1990. "Antibacterial properties of saliva: Role in maternal periparturient grooming and in licking wounds." *Physiology and Behavior* 48(3): 383–386.

Lukas, Dieter, and Tim Clutton-Brock. 2017. "Climate and the distribution of cooperative breeding in mammals." *Royal Society Open Science*. doi.org/10.1098/rsos.160897.

Sands, Jennifer, and Scott Creel. 2004. "Social dominance, aggression and faecal glucocorticoid levels in a wild population of wolves, Canis lupus." *Animal Behaviour* 67: 387–396.

Stahler, Daniel R., Douglas W. Smith, and Robert Landis. 2002. "The acceptance of a new breeding male into a wild wolf pack." *Canadian Journal of Zoology* 80: 360–365.

Thurston, Linda. 2002. "Homesite attendance as a measure of alloparental and parental care by gray wolves (*Canis lupus*) in northern Yellowstone National Park." Master's thesis, Texas A&M University.

Wilmers, Christopher C., Robert L. Crabtree, Douglas W. Smith, Kerry M. Murphy, and Wayne M. Getz. 2003. "Tropic facilitation by introduced top predators: Grey wolf subsidies to scavengers in Yellowstone National Park." *Journal of Animal Ecology* 72: 900–916.

Wilmers, Christopher C., Daniel R. Stahler, Robert L. Crabtree, Douglas W. Smith, and Wayne M. Getz. 2003. "Resource dispersion and consumer dominance: Scavenging at wolf- and hunter-killed carcasses in Greater Yellowstone, USA." *Ecology Letters* 6: 996–1003.

Yellowstone National Park. 2017. *Yellowstone Resources and Issues Handbook, 2017.* U.S. Department of the Interior, National Park Service, Yellowstone National Park.

Suggested Further Reading

Ferguson, Gary. 1996. *The Yellowstone Wolves: The First Year.* Helena, MT: Falcon Press.

Fischer, Hank. 1995. *Wolf Wars: The Remarkable Inside Story of the Restoration of Wolves to Yellowstone.* Helena, MT: Falcon Press.

Haber, Gordon, and Marybeth Holleman. 2013. *Among Wolves: Gordon Haber's Insights into Alaska's Most Misunderstood Animal.* Fairbanks, AK: Snowy Owl Books.

Halfpenny, James. 2012. *Charting Yellowstone Wolves: A Record of Wolf Restoration.* Gardiner, MT: A Naturalist's World.

Lopez, Barry Holstun. 1978. *Of Wolves and Men.* New York: Charles Scribner's Sons.

McIntyre, Rick. 1993. *A Society of Wolves: National Parks and the Battle Over the Wolf.* Stillwater, MN: Voyageur Press.

McIntyre, Rick. 1995. *War Against the Wolf: America's Campaign to Exterminate the Wolf.* Stillwater, MN: Voyageur Press.

McNamee, Thomas. 1997. *The Return of the Wolf to Yellowstone.* New York: Henry Holt.

McNamee, Thomas. 2014. *The Killing of Wolf Number Ten: The True Story.* Westport, CT: Prospecta Press.

Mech, L. David. 1981. *The Wolf: The Ecology and Behavior of an Endangered Species.* Minneapolis: University of Minnesota Press.

Mech, L. David, and Luigi Boitani, eds. 2003. *Wolves: Behavior, Ecology, and Conservation.* Chicago: University of Chicago Press.

Mech, L. David, Douglas W. Smith, and Daniel R. MacNulty. 2015. *Wolves on the Hunt: The Behavior of Wolves Hunting Wild Prey.* Chicago: University of Chicago Press.

Murie, Adolph. 1944. *The Wolves of Mount McKinley.* Washington, DC: United States Government Printing Office.

Phillips, Michael K., and Douglas W. Smith. 1996. *The Wolves of Yellowstone*. Stillwater, MN: Voyageur Press.

Phillips, Michael K., and Douglas W. Smith. 1997. *Yellowstone Wolf Project: Biennial Report 1995 and 1996*. National Park Service, Yellowstone Center for Resources, Yellowstone National Park.

Schullery, Paul. 1996. *The Yellowstone Wolf: A Guide and Sourcebook*. Worland, WY: High Plains Publishing.

Smith, Douglas W. 1998. *Yellowstone Wolf Project: Annual Report, 1997*. National Park Service, Yellowstone Center for Resources, Yellowstone National Park.

Smith, Douglas W., Kerry M. Murphy, and Debra S. Guernsey. 1999. *Yellowstone Wolf Project: Annual Report, 1998*. National Park Service, Yellowstone Center for Resources, Yellowstone National Park.

Smith, Douglas W., Kerry M. Murphy, and Debra S. Guernsey. 2000. *Yellowstone Wolf Project: Annual Report, 1999*. National Park Service, Yellowstone Center for Resources, Yellowstone National Park.

Smith, Douglas W., Kerry M. Murphy, and Debra S. Guernsey. 2001. *Yellowstone Wolf Project: Annual Report, 2000*. National Park Service, Yellowstone Center for Resources, Yellowstone National Park.

Smith, Douglas W., and Gary Ferguson. 2005. *Decade of the Wolf: Returning the Wild to Yellowstone*. Guilford, CT: Lyons Press.

Smith, Douglas W., ed. 2016. *Yellowstone Science: Celebrating 20 Years of Wolves* 24(1). National Park Service, Yellowstone Center for Resources, Yellowstone National Park.

Yellowstone National Park. 1997. *Yellowstone's Northern Range: Complexity and Change in a Wildland Ecosystem,* National Park Service, Mammoth Hot Springs, Wyoming.

Recommended DVD Documentaries

Catch Me If You Can II. 2016. Trailwood Films.

Wolf Pack. 2002. Trailwood Films.

Wolves: A Legend Returns to Yellowstone. 2000. National Geographic.

Further Information on the Wolves of Yellowstone

www.nps.gov/yell/learn/nature/wolves.htm

www.nps.gov/yell/learn/nature/wolfreports.htm

INDEX

Creek pack, 32–33; by Druid
Peak pack, 93, 146–47, 153,
155–56, 159–60, 170–71,
186–87, 201–3, 203–4, 210,
216–17, 218, 227–28, 234–
35, 242, 253; king of the
mountain, 228; by Leopold
pack, 89–92, 115; by Rose
Creek pack, 77–78, 94; skills
learned from, 33, 69, 89,
202–3, 216–17; snow sliding,
94; sparring, 89; standoff,
169; starting age, 201; stealing, 227–28; tackling, 210;
tossing, 32, 77–78, 90; tug
of war, 32, 89; wrestling,
84–86, 166, 201, 206–7
pregnancy, pseudo (false), 118
prey search image, 21
pronghorn, 66, 160–61, 215
pups, 25–26, 58, 64–65, 117–18.
See also play

radio collars, 25, 38–39, plate 8
rendezvous site, 89
Rose Creek wolf pack: 1997 litters, 98; 1999 litters, 183;
background, 14–15; care and
protection of litters, 99–101,
103–4, 105; communal den-
ning, 118; and Druid Peak
pack, confrontations with,
81–84, 98, 136, 228–30,
255–60, 261–62; and Druid

Peak pack, tensions and
territorial overlap, 121–22,
123–24, 129, 179, 231, 234,
236, 240–41, 251–52, 253–
54; elk hunting, 79, 80–81,
105, 107, 183, 244–47; evic-
tion and change in alpha
female, 232; family tree, xiv;
grizzly bear encounters, 77,
128–29; integration of wolf
8 as alpha male, 45–46, 47,
48, plate 3; membership
and growth, 93–94, 180–81;
move outside Yellowstone,
25; play, 77–78, 94; public
interest in, 40–41; release
into wild, 16–17, plate 1;
return to acclimation pen,
26–27, 37–38, 39–40
—wolf 7. *See under* Leopold wolf
pack
—wolf 8 (alpha male): achieve-
ments, 180–81, 254; aging,
183, 247–49; background,
15; breeding by, 55, 97, 98,
140, 232; bullying by Crystal
Creek siblings, 28–29; com-
parisons to wolf 21 joining
Druid Peak pack, 134–35;
confrontation with coyote
over food, 252; confron-
tation with Crystal Creek
pack, 48–50; confrontation
with Druid Peak pack at